P9-CCB-908

METHODS IN INOSITIDE RESEARCH

Methods in Inositide Research

Editor

Robin F. Irvine, M.A. (Biochem) Oxon, Ph.D. (Cantab)

AFRC Institute of Animal Physiology
and Genetics Research
Babraham, Cambridge, United Kingdom

Raven Press New York

Raven Press Ltd., 1185 Avenue of the Americas, New York, New York 10036

Made in the United States of America

Library of Congress Cataloging-in-Publication Data

Methods in inositide research / editor, Robin F. Irvine.
p. cm.
Includes bibliographical references.
Includes index.
ISBN 0-88167-677-2
1. Inositol phosphates. I. Irvine, Robin F.
[DNLM: 1. Biochemistry—methods. 2. Inositol Phosphates—analysis. QU 87 M592]
QP772.I5M48 1990
591.19'26—dc20
DNLM/DLC
for Library of Congress 90-8695
CIP

The material contained in this volume was submitted as previously unpublished material, except in the instances in which credit has been given to the source from which some of the illustrative material was derived.

Great care has been taken to maintain the accuracy of the information contained in the volume. However, neither Raven Press nor the editors can be held responsible for errors or for any consequences arising from the use of the information contained herein.

9 8 7 6 5 4 3 2 1

To my friend and mentor
Rex Dawson
a pioneer in inositide research

Preface

It is self-evident that interest in inositides has increased – exploded – over the last eight to nine years. What was previously a rather esoteric and obscure biochemical phenomenon of uncertain significance has become an enormously important intracellular signalling pathway. Its importance is more than matched by its complexity, as at least three second messengers are produced (diacylglycerol, inositol 1,4,5-trisphosphate and inositol 1,3,4,5-tetrakisphosphate) and, on top of this, many other inositol phosphates are synthesized and degraded by both resting and stimulated cells. Even diacylglycerol is no longer simple: we now realize that its stimulated production is not solely from inositides. These complexities necessitate (indeed, they are frequently revealed by) more complex and refined methods of analysis, and their proliferation is daunting to a seasoned inositide researcher let alone to a newcomer to the field.

As is described in the Introduction, only a few years ago the basic methods and the beginnings of sophistication were all capable of being summarized in a single chapter, but now the available methods are beyond the scope of one chapter or the personal experience of one person. This book is the result, but the philosophy remains the same. This is not a collection of comprehensive reviews nor of detailed discussion of principles. It is a selection of practical guides to methods that work, written by people, from a wide range of countries, who invented, refined, or use routinely the methods described. Often these practical guides are accompanied by an application to a biological system, to give a flavour of how the system works in practice. This is, therefore, I hope, a book to be held in the left hand while stirring the beaker with the right [left-handers are advised to adhere to this instruction exactly, to ensure preparation of D- and not L-enantiomers of inositol phosphates]. If more detailed discussion, history or theory is required, then a search in the literature cited herein is needed. Not every available method is covered, although many not described in detail are summarized in the Introduction, but I have tried to include at least one example of each type of analysis.

I am very grateful to Jasna Markovac of Raven Press for helping me talk myself into this project, to all the authors of the chapters for their prompt and enthusiastic response to the very minimum of bullying, to Stuart Laurie for ministering to the illustrations, and mostly to my wife, Sandi, for the enormous amount of work she has put into the organization, editing and overall quality of the book.

R. F. Irvine

February 1990 *Babraham, Cambridge*

Contents

Contributors

Kurt R. Auger *Department of Physiology, Tufts University School of Medicine, Boston, Massachusetts 02111*

Christilla Bachelot *Inserm U150, Hôpital Lariboisière, 75475 Paris, France*

Gerhard Bergmann *Institute for Analytical Chemistry, Faculty of Chemistry, Ruhr-University Bochum, D-4630 Bochum 1, FRG*

William B. Busa *Department of Biology, The Johns Hopkins University, Baltimore, Maryland 21218*

Lewis C. Cantley *Department of Physiology, Tufts University School of Medicine, Boston, Massachusetts 02111*

Steven Carter *Division of Molecular Biology and Biochemistry, School of Basic Life Sciences, University of Missouri-Kansas City, Kansas City, Missouri 64110*

Nullin Divecha *Department of Biochemistry, AFRC Institute of Animal Physiology and Genetics Research, Babraham, Cambridge CB2 4AT, UK*

Philippe Gascard *Physiologie de la Nutrition, Université Paris XI, Bâtiment 447, CNRS UA 646, 91045 Orsay Cedex, France*

Françoise Giraud *Physiologie de la Nutrition, Université Paris XI, Bâtiment 447, CNRS UA 646, 91045 Orsay Cedex, France*

Robin F. Irvine *Department of Biochemistry, AFRC Institute of Animal Physiology and Genetics Research, Babraham, Cambridge CB2 4AT, UK*

Warren G. King *Department of Biochemistry, University of Vermont College of Medicine, Burlington, Vermont 05405*

Arnis Kuksis *Banting and Best Department of Medical Research, University of Toronto, Toronto, Ontario, Canada M5G 1L6*

Andrew J. Letcher *Department of Biochemistry, Institute of Animal Physiology and Genetics Research, Babraham, Cambridge CB2 4AT, UK*

Teng-Nan Lin *Department of Biochemistry, School of Medicine, University of Missouri, Columbia, Missouri 65212*

Martin G. Low *Rover Physiology Research Laboratories, Department of Physiology and Cellular Biophysics, College of Physicians and Surgeons of Columbia University, New York, New York 10032*

J. Paul Luzio *Department of Clinical Biochemistry, University of Cambridge, Addenbrooke's Hospital, Cambridge CB2 2QR, UK*

Ronald A. MacQuarrie *Dionex Corporation, Sunnyvale, California 94086*

John A. Maslanski *Department of Biology, The Johns Hopkins University, Baltimore, Maryland 21218*

Georg W. Mayr *Institute for Physiological Chemistry, Faculty of Medicine, Ruhr-University Bochum, D-4630 Bochum 1, FRG*

John J. Myher *Banting and Best Department of Medical Research, University of Toronto, Toronto, Ontario, Canada M5G 1L6*

Susan Palmer *Molecular Pharmacology Group, Department of Biochemistry, University of Glasgow, Glasgow G12 8QQ, UK*

Noel Premkumar *Mobay Chemical Company, Stanley, Kansas 66085*

Susan E. Rittenhouse *Department of Biochemistry, University of Vermont College of Medicine, Burlington, Vermont 05405*

Peter Scholz *Institute for Analytical Chemistry, Faculty of Chemistry, Ruhr-University Bochum, D-4630 Bochum 1, FRG*

Leslie A. Serunian *Department of Physiology, Tufts University School of Medicine, Boston, Massachusetts 02111*

Gerry Smith *Department of Biochemistry, University of Cambridge, Cambridge CB2 1QW, UK*

Catherine E. L. Spencer *Department of Biochemistry, AFRC Institute of Animal Physiology and Genetics Research, Babraham, Cambridge CB2 4AT, UK*

Janet M. Stein *Department of Clinical Biochemistry, University of Cambridge, Addenbrooke's Hospital, Cambridge CB2 2QR, UK*

Leonard R. Stephens *Department of Biochemistry, AFRC Institute of Animal Physiology and Genetics Research, Babraham, Cambridge CB2 4AT, UK*

Jean-Claude Sulpice *Physiologie de la Nutrition, Université Paris XI, Bâtiment 447, CNRS UA 646, 91045 Orsay Cedex, France*

Grace Y. Sun *Department of Biochemistry, School of Medicine, University of Missouri, Columbia, Missouri 65212*

Michael J. O. Wakelam *Molecular Pharmacology Group, Department of Biochemistry, University of Glasgow, Glasgow G12 8QQ, UK*

Glossary

Individual abbreviations for the different inositol, phosphatidylinositol, and glycerophosphatidylinositol phosphates are not given below but the numbering is indicated by numbers in parentheses before the phosphate abbreviation in each case in the text: e.g., Ins(4)*P*, inositol 4-phosphate.[1]

ADP adenosine diphosphate
AMP adenosine monophosphate
ATP adenosine triphosphate
BHT butylated hydroxy toluene
BSA bovine serum albumin
CI chemical ionization
CL cardiolipin
COSY (one- or two-dimensional) correlated spectroscopy
c.p.m. counts per minute
CSF1 colony-stimulating factor 1
D deuterium (^2H)
DAG diacylglycerol
DEAE-cellulose diethylaminoethylcellulose
DMA dimethylacetals
DMEM Dulbecco's Modified Eagle's Medium
DMF *N,N*-dimethylformamide
DNPU dinitrophenylurethane
DPG bisphosphoglyceric acid
d.p.m. disintegrations per minute
EDTA ethylenediaminetetra-acetate
EGTA ethyleneglycolbis(aminoethylether)tetra-acetate
FAB fast atom bombardment
FAME fatty acid methyl esters
FFA free fatty acid
FID flame ionization detection
GC gas chromatography

---

[1] The numbering of inositol phosphate isomers throughout this book follows the recent recommendations of the Nomenclature Committee of the IUB (Biochem J 1989;258:1–2). Under these recommendations the IUPAC–IUB ruling about lowest-locant numbering may be relaxed if this clarifies metabolic interrelationships. Thus, for example, the product of 1-dephosphorylation of D-*myo*-inositol 1,3,4,5-tetrakisphosphate may be called D-*myo*-inositol 3,4,5-trisphosphate rather than L-*myo*-inositol 1,5,6-trisphosphate. The IUB also recommends the use of Ins as an abbreviation for D-*myo*-inositol. This recommendation is also followed without further comment, unless it is specifically stated otherwise.

GC-MS gas chromatography–mass spectrometry
Gro*P* glycerophosphate
Gro*P*Ins glycerophosphoinositol
Gro*P*Ins*P* glycerophosphoinositol monophosphate
Gro*P*InsP_2 glycerophosphoinositol bisphosphate
GTP guanosine triphosphate
Hepes 4(2-hydroxyethyl)-1-piperazinepropanesulphonic acid
HPLC high-performance liquid chromatography
HPTLC high-performance thin-layer chromatography
IDH *myo*-inositol dehydrogenase
Ins*P* inositol monophosphate
InsP_2 inositol bisphosphate
InsP_3 inositol trisphosphate
InsP_4 inositol tetrakisphosphate
InsP_5 inositol pentakisphosphate
InsP_6 inositol hexakisphosphate
LC liquid chromatography
MDD metal–dye detection
Mes 4-morpholine-ethane sulphonic acid
MIKES mass analysed ion kinetic spectra
NAD nicotinamide-adenine dinucleotide
NADH reduced nicotinamide-adenine dinucleotide
NCI negative chemical ionization
NMR nuclear magnetic resonance
P_i inorganic phosphate
PAD pulsed amperometric detection
PAR 4-(2-pyridylazo)resorcinol
PBS phosphate-buffered saline
PCA perchloric acid
PCI positive chemical ionization
PDGF platelet-derived growth factor
PEI polyethyleneimine
PEP phosphoenol pyruvate
PGE_1 prostaglandin E_1
PI phosphoinositide
PI 3-kinase phosphoinositide 3⁻kinase
PmlEtn ethanolamine plasmalogens
PMSF phenylmethylsulphonylfluoride
PP_i pyrophosphate
p.p.m. parts per million
PRP platelet-rich plasma
PtdCho phosphatidylcholine
PtdEtn phosphatidylethanolamine
PtdIns phosphatidylinositol

PtdIns*P* phosphatidylinositol monophosphate
PtdIns*P*$_2$ phosphatidylinositol bisphosphate
PtdOH phosphatidic acid
PtdSer phosphatidylserine
r.p.m. revolutions per minute
SAX strong anion exchange
S.E.M. standard error of the mean
SM sphingomyelin
TBA tetrabutylammonium
TBAP tetrabutylammonium phosphate
TBAHS tetrabutylammonium hydrogensulphate
TBDMS t-butyldimethylsilyl
TCA trichloroacetic acid
TEA tetraethylamine
TEAB triethylammonium bicarbonate
TID 3-(trifluoromethyl)-3-(*m*-iodophenyl)diazirine
TLC thin-layer chromatography
TMCS trimethylchlorosilane
TMS (1) trimethylsulphoxide, (2) trimethylsilyl
Tris 2-amino-2-hydroxymethylpropane-1,3-diol
UV ultraviolet
VIS visible light
WAX weak anion exchange

METHODS IN INOSITIDE RESEARCH

Methods in Inositide Research,
edited by Robin F. Irvine.
Raven Press, Ltd., New York © 1990

1

Introduction, and Survey of Other Methods

Robin F. Irvine

Department of Biochemistry, AFRC Institute of Animal Physiology and Genetics Research, Babraham, Cambridge CB2 4AT, UK

One Ring to rule them all, One Ring to find them
J. R. R. Tolkien, *The Lord of the Rings*

As this is very much a "cook-book," i.e., a collection of practical chapters rather than a series of extensive literature reviews, there is no intention of it being comprehensive. The philosophy behind this collection is identical with that behind a "Methods" chapter in *Phosphoinositides and receptor mechanisms* ((1); in those days it could be brief and still fairly comprehensive!). Methods are offered not necessarily because they are the best, but because they are tried and tested, and they work. What is best depends very much on what you want to do, and the tissue with which you want to do it, and no method will necessarily be ideal for every purpose. In this introductory chapter I shall summarize, or draw attention to, a few other methods which are not described in this book. Their omission occurred either because they are now well known, or because they are fully described in the original publications and need no detailed re-iteration here.

INOSITOL LIPIDS

Inositol lipids are assayed these days less frequently than are inositol phosphates or diacylglycerols (the "business end," of inositide signalling), not least because the homoeostatic mechanisms controlling polyphosphoinositol lipid synthesis mean that a stimulation of phosphoinositidase C is usually matched by a compensatory stimulation of the phosphatidylinositol (PtdIns) kinase and phosphatidylinositol 4-phosphate (PtdIns(4)*P*) kinase activities (and/or a decrease in phosphomonoesterase activities). Therefore, even if phosphatidylinositol bisphosphate (PtdInsP_2) mass levels do drop when the cell is stimulated, they soon return to (or exceed) the control level (see e.g., (2) and (3)). However, the turnover of the polyphosphoinositol lipids obviously increases, so isotopic incorporation under non-equilibrium conditions will change, and this is still a useful marker as an adjunct or alternative to inositol phosphate measurements (especially if the tissue being studied incorporates [^{3}H]inositol poorly). There are many thin-layer chromatography (TLC) methods which separate these lipids from each other and from other cellular lipids (e.g., see (1) and Chapter 13). Also, deacylation of the lipids followed by analysis of the glycerophosphoester "backbones" by Dowex anion exchange (4) or by high-performance liquid chromatography (HPLC) (see e.g., (5)) is an easy alternative, though obviously it is important to achieve complete removal of water-soluble radioactivity before deacylation as, for example, ATP will chromatograph on anion exchange close to glycerophosphoinositol bisphosphates (Gro*P*InsP_2). The more sophisticated method of analysis of deacylation and HPLC is at its most useful if the more recently recognized inositol lipids, e.g., PtdIns(3)*P*s, are the object of study (see Chapter 14).

Removal of the fatty acids (deacylation) is also desirable if the radioactivity in the individual phosphate moieties of PtdIns(4,5)P_2 are to be quantified (see (6) for the method of analysis). Note that, as has been stressed before (1), the deacylation method of Clarke and Dawson (7) is the method of choice for deacylation of lipids, because the removal of the fatty acids (as their methyl esters) is easy, and also because it minimizes the hydrolysis of the glycerophosphoester backbone. If such a hydrolysis occurs when the inositol lipids are analysed, the result is an artefactual production of inositol phosphates. Monomethylamine deacylation (7) minimizes this artefact, but it does not totally eliminate it, so if there is a great excess of one inositol lipid over another then it is essential to identify all the peaks on a resulting HPLC trace by using internal standards. For example, in higher plant tissues, where PtdIns(4,5)P_2 is at extremely low levels (8,9), it would be all too easy to identify some Ins*P*, derived artefactually from PtdIns, as Gro*P*Ins*P* (and consequently the Gro*P*Ins*P* as GroPInsP_2), and this has probably happened in some instances. Cutting the incubation time with the monomethylamine reagent down to 20 min will reduce the danger of splitting of the backbone, but will still fully deacylate most samples. Plant tissues also illustrate the potential dangers of using just TLC for identification of inositol lipids, as some plants contain lipids which incorporate inositol and ^{32}P, and which cochromatograph with PtdInsP_2, yet do not yield Gro*P*InsP_2, i.e., the lipids are not PtdInsP_2 (10). Notwithstanding these caveats, methylamine deacylation followed by Dowex chromatography (4) or straight TLC of the lipids ((1) and Chapter 13) are adequate for the great majority of tissues and purposes.

TLC is made much easier by the inclusion of internal standards of inositol lipids, and Low (Chapter 12) describes the preparation and analysis of polyphosphoinositol lipids by DEAE-cellulose and HPLC. The alternative preparative technique is a neomycin affinity column (originally described by Schacht (11) and refined by Palmer (12)). This is an excellent way of preparing pure lipids, but it is essential to analyse eluants by TLC (locating the lipids on the TLC plate by a phosphorus-detecting spray (1)) to characterize each batch of glass-neomycin beads. We have found that simply using the eluants described by Palmer (12) results in cross-contamination between the three principal inositol lipids, and to prepare really pure PtdIns*P*, e.g., for making high specific activity [^{32}P]Ins(1,4,5)P_3 (see Chapter 3), we use a stepwise elution from 200 mM to 500 mM in 100 mM steps (each batch of glass beads varies) and then take only the PtdIns*P* revealed to be pure by TLC analysis.

INOSITOL PHOSPHATES

Herein lies the main proliferation and analysis problem. Inositol phosphates exist in higher phosphorylated forms (e.g., inositol pentakisphosphates and hexakisphosphate (InsP_5/InsP_6)), whose metabolism is mostly separate from that of the "signalling" inositol phosphates, but is nevertheless connected to it (13,14). Yet it is an essential part of many studies to examine inositol 1,4,5-trisphosphate (Ins(1,4,5)P_3) and 1,3,4,5-tetrakisphosphate (Ins(1,3,4,5)P_4) in isolation, sometimes by mass, and sometimes by various radio-labelling techniques (and if the cells do not incorporate inositol readily, ^{32}P will have to be used). Also, the study of the intermediary metabolism of the higher inositol phosphates and their metabolic precursors and products has become important in itself. It is for these reasons that more space in this book is devoted to analysis of inositol phosphates than to anything else. A recent review by Dean and Beavan (15) summarizes some of the recent analytical methods, but here I want to draw attention to a few methods which are neither well-established nor described by other chapters in this book, but which we have found to be useful or to have promise. Choosing the appropriate method is not something to be done lightly, and I would not try to recommend one method over another – it depends on what is available by way of equipment and finance, and on what questions you are trying to ask. The only two sage pieces of advice I would offer are: (i) use the simplest and cheapest means adequate to answer the biological question you are asking; and (ii) there is never a substitute for internal standards, so use them whenever possible.

A powerful separation method for inositol phosphates, not described here, is iontophoresis in pyridine acetic acid (16), ammonium carbonate (7), or sodium oxalate (17). Practical details of their use are given elsewhere (1); note that the last buffer has been adapted for cellulose plates and bench-top electrophoretic apparatus (18). A very good separation of inositol phosphates and nucleotides is achieved with iontophoresis, and it is therefore suitable for analysing ^{32}P-labelled tissues (19,20). A problem of which to beware when analysing ^{32}P-labelled samples, or when measuring mass levels, is that there may be compounds lurking in the extract which are not inositol phosphates, but which purify with them, even through several stages (see e.g., (20)).

A simple separation of inositol phosphates (21) has also been achieved on Am-Preps (pre-packed anion exchange capsules sold by Amersham (UK)) which has the advantages of the earlier Sep-Pak separation (22). These advantages were speed, small elution volumes, lower ionic strength elutions and amenability to automation – a facility now realized by a 10-channel peristaltic pump (21). Unfortunately, the Sep-Paks' configuration was changed by the manufacturers so they no longer work as described (see Chapter 9). The Am-Preps are not a perfect substitute as, unlike the old Sep-Paks, they cannot be used more than once to achieve a really good separation of inositol phosphates, but although therefore expensive, they may be useful because they work extremely quickly. However, the innovation of isocratic HPLC using eluates from the standard Dowex columns (21,23) means that the two-stage analysis philosophy (21,22,23) no longer depends on these pre-packed columns for its exploitation. Maslanski and Busa (Chapter 9) describe a very promising use of Sep-Paks for separating Ins*P*s with TEAB, and it is to be hoped that the properties of these particular columns remain unchanged.

We do highly recommend the two-stage analysis philosophy (21,22,23); a rapid preliminary analysis by simple anion exchange ensures that the experiment has worked (i.e., good isotope incorporation, significant stimulation, close replicates, etc.) and it does not prevent a subsequent and easy analysis of isomers by automated isocratic HPLC (21,23). We find it much cheaper in time and consumables than detailed isomer analysis of the entire experiment by long gradient HPLC programs.

Other HPLC ion exchange methods not described in this book include citrate gradients on strong anion exchange columns (24) or sulphate on Mono-Q (25), and there is also a very sensitive technique of detection of inositol lipids and phosphates by malachite green (26), which may prove useful. Finally, we have found very useful (in the right circumstances) an elegant way of analysing inositol trisphosphate isomers which does not use HPLC. This is the method of Kennedy et al. (27), which rests in principle upon the fact that a rat brain supernatant, when incubated with Ins(1,3,4)P_3 in the presence of EDTA, will dephosphorylate it all to InsP_2s, while leaving Ins(1,4,5)P_3 untouched. Thus a sample can be split in half, and one half treated with the supernatant in this way. If both halves are then analysed by standard Dowex anion exchange chromatography, the Ins(1,3,4)P_3 and Ins(1,4,5)P_3 are therefore quantified by the difference. We found it advisable to omit Mg^{2+} altogether from the homogenization buffer (even a squeak of Mg^{2+} can activate the Ins(1,4,5)P_3 5-phosphatase by a small amount), and have had success in freezing the enzyme for storage at −15°C. We have also used a bovine brain supernatant and find that that works very well in an identical assay, though Ins(1,4,5)P_3 can be degraded by as much as 10% over 2 h at 37°C, so we always include a ^{32}P-labelled Ins(1,4,5)P_3 internal standard to correct for that (C. Spencer and R. Irvine, unpublished data).

This method depends for its success on there being only two InsP_3s present, and not much in the way of labelled higher inositol phosphates; Ins(1,3,4,5)P_4 is stable, but other InsP_4s, InsP_5 and InsP_6 are of unknown behaviour when incubated with brain supernatants in this way – they may give rise to other InsP_3s, which would complicate the analysis. However, in short-term labelled tissues (typically, tissue slices incubated with [^{3}H]inositol for a few hours) this is not a problem and so this method is a very cheap and easy assay for quantifying the two second messengers (Ins(1,4,5)P_3 and Ins(1,3,4,5)P_4) without using HPLC. In this context, it is of course self-evident that, if all that is required is to study the two known second messengers, then a specific mass assay using a binding protein or antibody must be considered as being near ideal. In Chapter 10, Palmer and Wakelam describe their assay for Ins(1,4,5)P_3; Donie and Reiser (28) have recently published a similar assay for Ins(1,3,4,5)P_4 that employs the Ins(1,3,4,5)P_4 "receptor" in cerebellum, and holds great promise.

DIACYLGLYCEROLS

Kuksis and Myher describe in full the mass quantification of diacylglycerol species including enantiomeric separations, and the detailed analysis of phospholipid radyl composition (see Chapter 17). A detailed mass analysis of this sort is becoming more important as data from many experimental systems reveal that, in a tissue where phosphoinositide turnover is increased, phosphoinositides are not necessarily either the only or the major source of diacylglycerol. If such detailed mass analysis is not required, an alternative is to label the tissue being studied to equilibrium with [^{3}H]glycerol, and then to analyse the diacylglycerols by the method of Itoh et al. (29) (i.e., using HPLC of the acetylated diacylglycerols). The fatty acid profile of the parent phospholipids can be deduced from the [^{3}H]glycerol-labelled extract by acetolysing (30) the phospholipid and then using the same HPLC technique. Patton et al. (31) have devised an alternative HPLC method in which the phospholipid species are separated directly without prior acetolysis. A suitable unlabelled standard to use for diacylglycerol analysis is from acetolysed hen egg phosphatidylcholine, and identification of peaks and uniformity of results is certainly helped by preparation of ^{14}C-labelled standards by acetolysis of ^{14}C-labelled phospholipids obtained from commercial sources.

For mass assays of total diacylglycerols, the assay of Preiss et al. (32), which employs a bacterial diacylglycerol kinase (obtained from Lipidex,

Westfield, NJ) is most widely used; it is reasonably straightforward and sensitive, and is specific for *sn*-1,2-diacylglycerols (as is protein kinase C, the principal target of diacylglycerol action and certainly the object of interest to most investigators). We have found that the source of cardiolipin necessary to activate the enzyme is crucial – some batches from various manufacturers were almost inactive, and in our experience Avanti (Birmingham, AL) have proved to be the most reliable source.

In Chapter 16 we describe an alternative assay which does not require cardiolipin using a diacylglycerol kinase prepared from rat brain. In our hands this is more sensitive, but the assay using bacterial enzyme is now in routine use in many laboratories because of its simplicity.

REFERENCES

1. Irvine RF. Biochemistry and analysis of inositides. In: Putney JW Jr, ed. *Phosphoinositides and receptor mechanisms*. New York: Alan R. Liss, 1986;89–107.
2. Perrett BP, Plantavid M, Chap H, Douste-Blazy L. Are polyphosphoinositides involved in platelet activation? *Biochem Biophys Res Commun* 1983;110:660–667.
3. Turner PR, Sheetz MP, Jaffe LR. Fertilization increases the polyphosphoinositide content of sea urchin eggs. *Nature* 1984;310:414–415.
4. Downes CP, Michell RH. Polyphosphoinositide phosphodiesterase of erythrocyte membranes. *Biochem J* 1981;198:133–140.
5. Stephens LR, Hawkins PT, Downes CP. Metabolic and structural evidence for the existence of a third species of polyphosphoinositide in cells: D-phosphatidyl-*myo*-inositol-3-phosphate. *Biochem J* 1989;259:267–276.
6. Hawkins PT, Michell RH, Kirk CG. Analysis of the metabolic turnover of the individual phosphate groups of phosphatidylinositol 4-phosphate and phosphatidylinositol 4,5-bisphosphate. *Biochem J* 1984;218:785–789.
7. Clarke NG, Dawson RMC. Alkaline $O \rightarrow N$-transacylation. A new method for the quantitative deacylation of phospholipids. *Biochem J* 1981;195:301–306.
8. Irvine RF, Letcher AJ, Lander DJ, Drøbak BK, Dawson AP, Musgrave A. Phosphatidylinositol(4,5)bisphosphate and phosphatidylinositol (4)phosphate in plant tissues. *Plant Physiol* 1989;89:888–892.
9. Coté CG, Depass AL, Quarmby LM, Tate BF, Morse MJ, Salter RL, Crain RC. Separation and characterization of inositol phospholipids from the pulvini of *Samanea saman*. *Plant Physiol* 1989;90:1422–1428.
10. Drøbak BK, Ferguson IB, Dawson AP, Irvine RF. Inositol containing lipids in suspension-cultured plant cells. *Plant Physiol* 1988;87:217–222.
11. Schacht J. Purification of polyphosphoinositides by chromatography on immobilized neomycin. *J Lipid Res* 1978;19:1063–1067.
12. Palmer FB. Chromatography of acidic phospholipids on immobilized neomycin. *J Lipid Res* 1981;22:1296–1300.
13. Irvine RF, Moor RM, Pollock WK, Smith PM, Wreggett KA Inositol phosphates: proliferation, metabolism and function. *Philos Trans Roy Soc Lond* [*Biol*] 1988;320:281–298.
14. Shears SB. Metabolism of the inositol phosphates produced upon receptor activation. *Biochem J* 1989;260:313–324.
15. Dean NM, Beavan MA. Methods for the analysis of inositol phosphates. *Anal Biochem* 1989;183:199–209.
16. Dawson RMC, Clarke NG. D-*myo*-inositol 1:2 cyclic phosphate 2-phosphohydrolase. *Biochem J* 1972;127:113–118.
17. Seiffert UB, Agranoff BW. Isolation and separation of inositol phosphates from hydrolysis of rat tissues. *Biochim Biophys Acta* 1965;98:574–581.
18. Whipps DE, Armston A, Pryor H, Halestrap AP. Effects of glucagon and Ca^{2+} on the

metabolism of phosphatidylinositol 4-phosphate and phosphatidylinositol 4,5-bisphosphate in isolated rat hepatocytes and plasma membranes. *Biochem J* 1987;241:835–845.
19. Agranoff BW, Murthy P, Seguin EB. Thrombin induced phosphodiesteric cleavage of phosphatidylinositol bisphosphate in human platelets. *J Biol Chem* 1983;258:2076–2078.
20. Harrison R, Rodan E, Lander D, Irvine RF. Ram spermatozoa produce inositol 1,4,5-trisphosphate but not inositol 1,3,4,5-tetrakisphosphate during the Ca^{2+}/ionophore induced acrosome reaction. *Cellular Signalling* 1990;2:273–284.
21. Wreggett KA, Lander DJ, Irvine RF. Semi-automated fractionation of isomers of inositol phosphates. *Methods in Enzymol* 1990; (in press).
22. Wreggett KA, Irvine RF. A rapid separation method for inositol phosphates and their isomers. *Biochem J* 1987;247:655–660.
23. Wreggett KA, Irvine RF. Automated isocratic high performance liquid chromatography of inositol phosphate isomers. *Biochem J* 1989;262:997–1000.
24. Matthews WR, Guido DM, Huff RM. Anion-exchange high-performance liquid chromatographic analysis of inositol phosphates. *Anal Biochem* 1988;168:63–70.
25. Meek JL. Inositol bis-, tris-, and tetrakis (phosphate)s: analysis in tissues by HPLC. *Proc Natl Acad Sci USA* 1986;85:4162–4166.
26. Underwood RH, Greeley R, Glennon ET, Menachery ATI, Braley LM, Williams GH. Mass determination of polyphosphoinositides and inositol trisphosphate in rat adrenal glomerulosa cells with a microspectrophotometric method. *Endocrinology* 1988;123:213–219.
27. Kennedy ED, Batty IR, Chilvers ER, Nahorski SR. A simple method to separate [^{3}H]inositol 1,4,5- and 1,3,4-trisphosphate isomers in tissue extracts. *Biochem J* 1989;260:283–286.
28. Donie F, Reiser G. A novel, specific binding protein assay for quantitation of intracellular inositol 1,3,4,5-tetrakisphosphate (InsP_4) using a high-affinity receptor from cerebellum. *FEBS Lett* 1989;254:155–158.
29. Itoh K, Suzuki A, Kuoki Y, Akino T. High performance liquid chromatographic separation of diacylglycerol acetates to quantitate disaturated species of lung phosphatidylcholine. *Lipids* 1985;20:611–616.
30. Choe H-G, Wiegand RD, Anderson RE. Quantitative analysis of retinal glycerolipid molecular species acetylated by acetolysis. *J Lipid Res* 1989;30:454–457.
31. Patton GM, Fasulo JM, Robins CJ. Separation of phospholipids and individual molecular species of phospholipids by high-performance liquid chromatography. *J Lipid Res* 1982;23:190–196.
32. Preiss J, Loomis CR, Bishop WR, Stein R, Niedel JE, Bell RM. Quantitative measurement of *sn*-1,2-diacylglycerols present in platelets, hepatocytes, and *ras*- and *sis*-transformed normal rat kidney cells. *J Biol Chem* 1986;261:8597–8600.

Methods in Inositide Research,
edited by Robin F. Irvine.
Raven Press, Ltd., New York © 1990

2

Preparation and Separation of Inositol Tetrakisphosphates and Inositol Pentakisphosphates and the Establishment of Enantiomeric Configurations by the Use of L-Iditol Dehydrogenase

Leonard R. Stephens

Department of Biochemistry, AFRC Institute of Animal Physiology and Genetics Research, Babraham, Cambridge CB2 4AT, UK

PREPARATION OF ^{32}P- AND ^{3}H-LABELLED INOSITOL TETRAKISPHOSPHATES

Three inositol tetrakisphosphates (InsP_4s) have been characterized in plant and animal cells: Ins(1,3,4,5)P_4, Ins(1,3,4,6)P_4 and Ins(3,4,5,6)P_4. [^{3}H]Ins(1,3,4,5)P_4 is now commercially available and several detailed descriptions of its preparation have been published (1).

Ins(1,3,4,6)P_4

[^{3}H]Ins(1,3,4,6)P_4 has been prepared from ^{3}H-labelled inositol 1,3,4-trisphosphate ([^{3}H]Ins(1,3,4)P_3; commercially available) by several groups (2,3,4) and although purified Ins(1,3,4)P_3 6-OH kinase preparations (4,5) are the enzyme sources of choice, reasonable yields (greater than 40% of starting material) can be obtained using a crude cytosol preparation from rat brain. However, if rat brain cytosol is used as a source of Ins(1,3,4)P_3 6-OH kinase activity, the product needs to be purified on a high-performance liquid chromatography (HPLC) column (see below), because approximately 5–10% of the total [^{3}H]InsP_4 recovered will be [^{3}H]Ins(1,3,4,5)P_4 as a consequence of an Ins(1,3,4)P_3 5-OH kinase activity found in rat brain cytosol [L. Stephens, unpublished data; (6)].

The reaction is run in 400 μl of buffer and we have found that this volume can be loaded on to a typical analytical-scale anion exchange HPLC column without adversely effecting column performance. The buffer contains 5 mM ATP, 6 mM $MgCl_2$, 10 mM creatine phosphate, 1 mM dithiothreitol, 100 mM KCl, 50 mM Hepes (pH 7.0, 37°C), 2 mM EGTA, 5 units creatine phosphokinase/ml, and 80 μl of a 100,000 *g* rat brain supernatant (prepared from one 250 g male rat brain homogenized in 5 ml of 0.25 M sucrose, 50 mM Hepes (pH 7.0, 4°C), 2 mM EGTA, 0.1 mM phenylmethylsulphonylfluoride, 1 mM dithiothreitol). Incubation is at 37°C for 20 min. It would be possible to prepare [^{32}P]Ins(1,3,4,6)P_4 from [^{32}P]Ins(1,3,4)P_3, by the above protocol but the specific radioactivity of the [^{32}P]Ins(1,3,4)P_3 would need to be substantial because the K_m of the Ins(1,3,4)P_3 6-OH kinase is very low (5,6). However, this strategy could yield [^{32}P]Ins(1,3,4,6)P_4 specifically labelled in the 3 position, and this would enable simple assays to be constructed to study the metabolism of the 3-phosphate in isolation. If [γ-^{32}P]ATP and "cold" Ins(1,3,4)P_3 were used as substrates, this would allow specific 6-OH labelling of Ins(1,3,4,6)P_4 to be achieved. Such preparations are always awkward because not only does the enzyme preparation need to be substantially purified (at the very least to remove contaminating ATPases) but the [γ-^{32}P]ATP would also need to be purified; a number of ^{32}P-labelled polyphosphates that accumulate in all commercially available preparations of [γ-^{32}P]ATP elute from anion exchange HPLC columns close to InsP_4 (4,7,8). Purification of the [γ-^{32}P]ATP is most effectively achieved by passing it through an anion exchange HPLC column immediately prior to use; any acidity in the eluting buffer needs to be neutralized before its inclusion in the assay (8). Alternatively, if a KCl or NaCl gradient is employed, the ATP fractions can be diluted directly into the assay buffer (9).

Ins(3,4,5,6)P_4

^{3}H- and ^{32}P-labelled Ins(3,4,5,6)P_4 have so far been successfully prepared only from appropriately radiolabelled intact cells. A very convenient cell type for this purpose is the chick erythrocyte, as it is easily prepared, robust and readily incorporates both [^{3}H]Ins and [^{32}P]P_i into inositol phosphates. Avian erythrocytes also contain Ins(1,3,4,5)P_4, Ins(1,3,4,6)P_4 and Ins(1,3,4,5,6)P_5 (see below), which also become rapidly labelled by exogenously supplied [^{32}P]P_i and [^{3}H]Ins. This has two implications: firstly, the Ins(3,4,5,6)P_4 needs to be purified by anion exchange HPLC to ensure that it is pure; secondly, the same avian erythrocyte preparations can be used to obtain supplies of [^{3}H]Ins or [^{32}P]P_i-labelled Ins(1,3,4,6)P_4, Ins(1,3,4,5)P_4, and Ins(1,3,4,5,6)P_5. Conditions for preparation, incubation with isotopes, extraction and purification have been published in detail elsewhere (4,7,10).

Points to note in these procedures are as follows. Heparin is the anticoagulant of choice as it lyses the majority of white cells, and if the first centrifugation step in the erythrocyte washing protocol is particularly hard (5–10 min at 3000–4000 *g*) the debris from these cells forms a congealed plug that can be very cleanly removed. If this is not done, avian white cells can be a problem. If the cells are to be labelled with [^{32}P]P_i, we incubate in a chloride-free medium (several mM Cl^- is sufficient to inhibit [^{32}P]P_i entry significantly) containing bovine serum albumin (BSA) and antibiotics (10). The cells are washed twice with the chloride-free buffer before the tracer is finally added.

We have found that an acid extract from approximately 0.3 ml of packed erythrocytes is all that can be effectively chromatographed on an analytical-scale Partisphere weak anion exchange (WAX) HPLC column. Hence incubations are best done in a total volume of 0.8–3.0 ml of medium containing 0.3 ml of packed cells. Larger volumes are necessary if the incubation is to continue for longer periods, e.g., 24 h. Our cell suspensions are usually incubated in a single 1 ml well of a typical 12 well tissue-culture plate, with the lid held in place by a strip of Parafilm (to reduce water loss through evaporation), and the entire plate is attached to the bottom of a large, flat-bottomed, glass trough (which is then covered over with Parafilm) using a sheet of Plasticene. The entire "Box of Delights" is then placed in a shaking, thermostatically controlled (38–40°C) water-bath. After 20–36 h the cells are removed, placed in 10 ml of ice-cold isotonic saline in a 10 ml pointed-tipped polypropylene test-tube, and pelleted by centrifugation. Well over 90% of the cells should be recovered. The saline supernatant should be rapidly aspirated and the cells lysed by (i) adding 1 ml of ice-cold H_2O, (ii) vortexing for 5 s, (iii) adding 50 µl of 70% (v/v) perchloric acid, and (iv) vortexing for 20–30 s. The cell debris produced during the above procedure is pelleted by centrifugation at the maximum setting of a bench-top centrifuge for 5 min (with the centrifuge either refrigerated at 4°C or placed in a cold room). The acid-soluble supernatants from [^3H]Ins-labelled extracts are then neutralized (see below) immediately. ^{32}P-labelled extracts are incubated for 15 min, on ice, with charcoal (see below) then neutralized.

Charcoal Extraction

If a charcoal extraction (using Norit GSX or Darco-G60) is to be incorporated into the purification procedure, to remove the majority of [^{32}P]ATP from cell extracts or assays, it should be thoroughly acid-washed (five times with ice-cold 5% perchloric acid). It should then be mixed only with inositol phosphate extracts under acidic conditions (e.g., in the presence of 4% ice-cold perchloric acid (10)) or at high salt concentrations (11). The charcoal is added as a 25 mg/ml slurry in 4% perchloric acid with an air displacement pipette (100 µl is used with an extract from 0.3 ml of packed cells). The charcoal is removed after 15 min (on ice) by adding 40 μl of 10% (w/v in H_2O) BSA, vortexing, and then centrifuging the samples.

NEUTRALIZATION OF PERCHLORIC ACID EXTRACTS

We invariably quench assays and cell preparations with ice-cold perchloric acid (final concentrations range from 2% to 5%). Perchloric acid has the advantage that a number of techniques can be employed that simultaneously remove the acid and neutralize the extracts, hence minimizing the ionic strength of the final sample. The two neutralization strategies we employ most frequently are KOH and tri-*n*-octylamine/Freon (1:1, v/v (12)). The former technique takes advantage of the relatively low solubility of potassium perchlorate. Typically we vortex a perchloric acid-precipitated suspension and then pellet the precipitated cellular or assay debris by centrifugation. The supernatant above the acid-insoluble material is removed and added directly to a previously established fixed volume of 2 M KOH, 0.1 M Mes buffer (pK_a around 6) and 10–20 mM EDTA, such that the final pH is ≈ 6.0. The advantage of this procedure is that in our experience absolutely no problems are encountered with non-specific losses of inositol polyphosphates despite (i) an absence of carrier, (ii) low ionic strength conditions or (iii) the presence of divalent cations (so long as sufficient EDTA has been added). The disadvantages are, in essence, the care which must be taken over the neutralization to ensure that the samples arrive at the correct final pH, and, the fact that the solubility of potassium perchlorate is not insignificant. The solubility of potassium perchlorate in H_2O is dependent on temperature and even at 0–4°C, potassium perchlorate solutions do not saturate until they reach concentrations of approximately 50 mM. This problem can be minimized only by performing the neutralization and subsequent centrifugation steps at 0–4°C and in a small total volume.

If tri-*n*-octylamine/Freon is to be used as the neutralizing agent, the ice-cold supernatant from the centrifugation of the perchloric-acid-quenched samples is added to 1.1 volumes of tri-*n*-octylamine/Freon (freshly prepared at room temperature) into which an appropriate portion of Na_2 EDTA (e.g., 50 μl of 0.1 M) solution has been placed. The sample is then very well mixed, and centrifuged at the maximum setting of a bench-top centrifuge or in a microfuge for 5–10 min (again at room temperature). A three-layered system should result: an upper aqueous phase, that should be pH 5–6 (depending on the pH of the original sample), a viscous middle layer of tri-*n*-octylamine perchlorate, and a lower phase containing the Freon and any excess tri-*n*-octylamine. The advantages of this approach are obvious: it is simple, rapid, and results in a low concentration of perchlorate in the upper phase. The disadvantages are: (a) that some organic residues invariably contaminate the upper phase and these may occasionally interfere with subsequent assays (e.g., Ins(1,4,5)P_3 binding assays (13); (b) that inositol polyphosphates (in our experience only inositol pentakisphosphates (InsP_5s) and hexakisphosphate (InsP_6)) can "get lost" in the tri-*n*-octylamine/perchlorate residue. This non-specific loss of inositol polyphosphates occurs most commonly in the following circumstances.

1. When the concentrations of InsP_5 or InsP_6 are very low. This can sometimes be alleviated by addition of a sample of phytic acid hydrolysate as carrier (14), although this is not always possible, e.g., if the specific radioactivity or mass of the inositol polyphosphates is to be measured or if the polyphosphates are destined to be metabolized by an enzyme.
2. When the ionic strength of the solutions is low, e.g., in acid extracts from small quantities of tissue (under these circumstances problem 1 above often also applies).
3. If any cellular debris from the initial precipitation is transferred into the neutralization vessel.

The various modifications that have been employed to circumvent these problems (e.g., pre-neutralization with Tris/maleate buffers (14)) are probably more trouble to sort out than simply changing to the KOH/Mes precipitation described.

SEPARATION AND PURIFICATION OF ^{3}H- AND ^{32}P-LABELLED InsP_4s

Separation of InsP_4s from Avian Erythrocytes on a Strong Anion Exchange (SAX) HPLC Column

A charcoal-treated acid-extract of ^{32}P-labelled avian erythrocytes still contains substantial quantities of ^{32}P-labelled nucleotide phosphates and [^{32}P]-P_i (see the previous section). These contaminating compounds are most effectively removed by applying the sample to a Partisil 10 SAX HPLC column and eluting with a rapidly developing ammonium formate/phosphoric acid gradient [i.e., conditions originally designed to separate Ins(1,3,4)P_3 and Ins(1,4,5)P_3 (15)]. By this means, Ins(1,3,4,5)P_4, Ins(1,3,4,6)P_4 and Ins(3,4,5,6)P_4 elute together but separate from InsP_5. The InsP_4 and InsP_5 peaks are then collected and desalted [see below and (10)].

Better resolution of InsP_4s can be obtained on SAX HPLC columns by using phosphate buffers (16). Under these conditions, InsP_4s elute from a standard analytical-scale SAX HPLC column with 1 M $NaH_2PO_4 \cdot NaOH$ (pH 3.8, 25°C) in the order Ins(3,4,5,6)P_4, Ins(1,3,4,5)P_4 and Ins(1,3,4,6)P_4 (L. Stephens, unpublished data). However, Ins(3,4,5,6)P_4 is usually poorly resolved from Ins(1,3,4,5)P_4 and, as the major [^{3}H]InsP_4 in extracts from control cells is usually [^{3}H]Ins(3,4,5,6)P_4 and the minor one [^{3}H]Ins(1,3,4,5)P_4, this is not an ideal system for analysing cell-derived extracts; WAX (described below) is more suitable.

"Other" InsP_4s Eluting on SAX HPLC Columns

D/L [^{3}H]- or [^{32}P]Ins(1,2,3,4)P_4 (in a roughly equal mixture of the two enantiomers) and [^{3}H]- or [^{32}P]Ins(1,2,5,6)P_4 can be obtained from mung bean seeds germinated in the dark with solutions of [^{3}H]Ins or [^{32}P]P_i and extracted with acid (see p. 16 for a description of preparation of [^{32}P]InsP_5s from mung beans). Both Ins(1,2,3,4)P_4 and Ins(1,2,5,6)P_4 elute from a Partisphere SAX column after Ins(1,3,4,5)P_4; Ins(1,2,5,6)P_4 is retained longer (L. Stephens, unpublished data).

D/L-[^{3}H]Ins(1,2,4,6)P_4 has been prepared from [^{3}H]Ins(1,3,4,6)P_4 by controlled acid-catalysed phosphate migration (at 80°C or 100°C for 8 min with 1 M HCl: L. Stephens, unpublished data; (8); it elutes from a Partisphere SAX column substantially later than Ins(1,3,4,6)P_4.

Purification of InsP_4s on WAX HPLC Columns

The preparations of Partisil 10 SAX-purified [^{32}P]InsP_4 standards (see above and desalted as described below), or alternatively, the crude, neutralized ^{3}H-labelled cell extracts, are then resolved on a Partisphere WAX HPLC column. These columns separate all three species of InsP_4 found in the avian erythrocytes [in order of increasing retention time: Ins(1,3,4,6)P_4, Ins(1,3,4,5)P_4 and Ins(3,4,5,6)P_4 (4)]. A new column is eluted isocratically with 0.16 M $(NH_4)_2HPO_4 \cdot H_3PO_4$ (pH 3.2, 22°C) and InsP_4s should emerge after 40–80 min. Once the column begins to age (after more than six runs) the elution times of the InsP_4s will have reduced to a point where they fail to resolve. At this point (or preferably just before!) the concentration of the eluting buffer is reduced [by approximately 10–15% initially; see (10) for further information]. Obviously the crude [^{3}H]Ins$P$ preparations contain a substantial amount of [^{3}H]Ins(1,3,4,5,6)P_5, and this is eluted from a new Partisphere WAX HPLC column by approximately 0.4 M $(NH_4)_2HPO_4 \cdot H_3PO_4$ (pH 3.2 at 22°C). The [^{3}H]InsP_4 peaks are identified by either: (i) spiking the samples with trace quantities of [^{32}P]InsP_4 "standards" (for preparation see above), which can then be detected in the column eluant by Čerenkov counting; or (ii) removing small portions from each of the fractions collected and counting them individually for ^{3}H.

DESALTING INOSITOL PHOSPHATE PREPARATIONS

Routinely, we desalt HPLC-purified inositol phosphates (in either ammonium formate/phosphoric acid or ammonium phosphate/phosphoric acid buffers (e.g., (7), (15) and (16)) by adjusting the pH of the buffer in which they were eluted to 6–7 with triethylamine, and diluting with H_2O (8- to 10-fold) before finally applying them to a column containing 100–200 μl of packed AG 1.8 200–400 resin (BioRad) in the formate form. The columns are washed with 6–8 ml of 0.2 M ammonium formate, 0.1 M formic acid (to elute P_i) and then with 2.5 ml of 2 M ammonium formate, 0.1 M formic acid to elute InsP_4s or InsP_5. The eluants are roughly neutralized with NH_4OH (approximately 28 μl of stock NH_4OH) and freeze-dried directly. We freeze-dry samples in two stages: first on a freeze-drier with a built-in centrifuge (e.g., a Savant) and then, once it is solid, in a standard "high-vacuum" freeze-drying system. Managed in this way the samples produced by the above procedure can be completely free of ammonium formate in about 15 h. If the inositol phosphates are carrier-free or present in small amounts (less than 5–10 nmol) they can become non-specifically bound to the surfaces on to which they are freeze-dried. This problem can be minimized by the use of polypropylene containers (although build up of static then becomes a major difficulty) or freshly siliconized glassware. The best strategy that we have tried for "rescuing" inositol phosphates that have stuck to glass or plastic is using dilute NH_4OH.

InsP_5s: PREPARATION OF ^{3}H- AND ^{32}P-LABELLED STANDARDS AND THEIR SEPARATION BY ANION EXCHANGE HPLC

There are six possible *myo*-inositol pentakisphosphates: two pairs of enantiomers (D- and L-Ins(1,2,4,5,6)P_5 and D- and L-Ins(1,2,3,4,5)P_5), and two isomers in which the phosphates do not disturb the plane of symmetry in the parent *myo*-inositol moiety (Ins(1,2,3,4,6)P_5 and Ins(1,3,4,5,6)P_5).

Ins(1,3,4,5,6)P_5

Ins(1,3,4,5,6)P_5 is the only isomer of InsP_5 that has been described in animal cells. Johnson and Tate (17) originally characterized the inositol polyphosphate extracted from avian erythrocytes as Ins(1,3,4,5,6)P_5 and these cells still remain the best source of this InsP_5 isomer either for labelling with ^{3}H or ^{32}P (see above) or for larger-scale preparations (17).

Ins(1,2,3,4,6)P_5

This has been detected in *Dictyostelium discoideum* (a slime mould; L. Stephens, unpublished data) but is probably best prepared (in either ^{3}H- or ^{32}P-labelled form, or on a large scale) by alkaline hydrolysis of InsP_6 (see below).

D/L-Ins(1,2,3,4,5)P_5

D- and/or L-Ins(1,2,3,4,5)P_5 is one of the principal types of InsP_5 that can be isolated chromatographically from germinating mung bean (*Phaseolus aureus*) seedlings. Typically, one mung bean is germinated in 0.75 ml of sterile H_2O containing [^{3}H]Ins or [^{32}P]P_i, in the dark, for 50–60 h at 25°C. The resulting seedling is homogenized in 2 ml of ice-cold 5% perchloric acid using a "studded" glass-on-glass hand homogenizer. The acid-insoluble debris is pelleted in a microfuge and the supernatant neutralized with tri-*n*-octylamine/Freon as described above. The InsP_5 can be purified directly from the acid extract by anion exchange HPLC (see below). Two InsP_5s can be isolated from mung bean seedlings: D/L-Ins(1,2,3,4,5)P_5 (proportion of enantiomers undetermined) and D/L-Ins(1,2,4,5,6)P_5 (largely the L-isomer; L. Stephens, unpublished data). The D/L-Ins(1,2,4,5,6)P_5, which is retained longer during HPLC on both SAX and WAX, has been completely characterized by ^{31}P and two-dimensional ^{1}H nuclear magnetic resonance (NMR) (T. Moore, R. Morris, L. Stephens, P.T. Hawkins and R.F. Irvine, unpublished data).

Ins(1,2,3,4,5)P_5, with either little or no L-Ins(1,2,3,4,5)P_5 in it, was determined to be the minor product of a fungal phytase (18) acting on InsP_6. Its enantiomeric configuration was determined by measuring the specific optical rotation of the product and comparing it to a characterized Ins(1,2,3,4,5)P_5 preparation. By incubating a commercially available preparation of this enzyme activity with ^{3}H- or ^{32}P-labelled InsP_6 (see below) it is therefore possible to prepare ^{3}H- or ^{32}P-labelled Ins(1,2,3,4,5)P_5.

The conditions we employ are: [^{3}H]- or [^{32}P]InsP_6 (at concentrations of about 3 μM, i.e., in the first-order range), 20 mM $CH_3COOH \cdot NaOH$ (pH 5.0, 37°C), 1 mg BSA/ml, 0.0008 units *Aspergillus* phytase/ml (units defined by Sigma, from whom it is obtained). Optimal yields of InsP_5s (the major isomer is Ins(1,2,4,5,6)P_5) are obtained after 60 min at 37°C (L. Stephens, unpublished data). The InsP_5 products can then be resolved by anion exchange HPLC (see below).

Ins(1,2,3,5,6)P_5 (= L-Ins(1,2,3,4,5)P_5) is the major product of a crude preparation of wheat-bran phytase (19) which has been more recently resolved both by differing properties (e.g., inhibitors, pH) and by chromatography. The F_1 component of this preparation is largely responsible for the specific removal of the 4-phosphate group (20). This pioneering work has been exploited to prepare ^{3}H- and ^{32}P-labelled Ins(1,2,4,5,6)P_5 from ^{3}H- or ^{32}P-labelled InsP_6 (see below; L. Stephens and P.T. Hawkins, unpublished data). The method is as follows: a commercially available, crude wheat-bran phytase (Sigma) is incubated in a buffer containing 20 mM $CH_3COOH \cdot NaOH$ (pH 5.0 at 37°C), 1 mg BSA/ml, 3 μM InsP_6 (first-order reaction conditions), 0.00072 wheat-bran phytase units/ml (units defined by Sigma). After 60 min, approximately 10% of the InsP_6 phosphorus will have been released (this is the degree of reaction that yields optimal amounts of InsP_5). The assay is quenched with perchloric acid and processed with KOH/Mes, as described above. The [^{3}H]- or [^{32}P]InsP_5 is purified by anion exchange HPLC and desalted (see above and below).

D/L-Ins(1,2,4,5,6)P_5

D -Ins(1,2,4,5,6)P_5 is the major product of a fungal phytase (18) and can be prepared with a commercially available crude preparation of *Aspergillus* phytase (Sigma; see above). D/L-Ins(1,2,4,5,6)P_5 is also usually the major product formed during the early stages of the alkaline hydrolysis of InsP_6 (see below) and so this technique yields the best source of racaemic mixtures of the two enantiomers.

PREPARATION OF [^{3}H]InsP$_6$ OR [^{32}P]InsP$_6$

[^{3}H]InsP_6 has recently become commercially available (NEN, 16 Ci/mmol), but it can be easily prepared at high specific radioactivity and yield by direct phosphorylation of inositol [(21); N. Carter, P.T. Hawkins and L. Stephens, unpublished data]. [^{3}H]Ins is dried down into a conical glass vial, then dissolved in 10 μl of orthophosphoric acid by heating to 60°C for 15 min; 0.18 g of polyphosphoric acid is added and the mixture is placed under vacuum and heated to 150°C. After 6 h, 0.5 ml of H_2O is added then the mixture is heated to 100°C for 3 h (no vacuum) and cooled. The reaction mixture is washed out with H_2O, diluted to 100 ml, the pH adjusted to 6.5 with NaOH and the sample applied to a small desalting column (as described above; the column is washed with 0.2 M ammonium formate, 0.1 M formic acid and then the InsP_6 is eluted with 2.0 M ammonium formate, 0.1 M formic acid). After the sample has been freeze-dried the [^{3}H]InsP_6 is purified by HPLC. A small portion of [^{32}P]InsP_6 (see below) can be used to identify the peak, but this will alter the specific radioactivity of the preparation from that of the starting preparation of [^{3}H]Ins. Typically, 20% of the starting [^{3}H]Ins will be recovered as [^{3}H]InsP_6. Some ^{3}H-labelled compounds elute after InsP_6; these have not yet been identified. [^{32}P]InsP_6 can be easily prepared (although not at very high specific radioactivity) from [^{32}P]P_i-labelled germinating mung beans. A bean is labelled and extracted as described for the preparation of [^{32}P]InsP_5s above, and the [^{32}P]InsP_6 is purified on an anion exchange HPLC column. The [^{32}P]InsP_6 can be readily identified as the most polar ^{32}P-labelled metabolite eluting from the column in significant quantities (a typical yield from one bean germinated with 0.5 mCi of [^{32}P]P_i for 50–60 h would be 15 μCi of [^{32}P]InsP_6).

ALKALINE HYDROLYSIS OF InsP$_6$

[^{3}H]InsP_6 or [^{32}P]InsP_6 can be dephosphorylated under alkaline conditions to yield a mixture of all the InsP_5s. Heating a solution of InsP_6 (in 50 mM $Na_2HPO_4 \cdot NaOH$, pH 10.5, 25°C) at 120°C in an autoclave for 3 h results in a fairly equal mixture of InsP_5s [in about 15–20% yield from the InsP_6 (22)]; however, it must be emphasized that the Ins(1,2,3,4,5)P_5 and Ins(1,2,4,5,6)P_5 species will be racaemic mixtures. The four chromatographically distinct types of InsP_5 can all be purified from the hydrolysate by anion exchange HPLC (see below) after the pH of the sample has been titrated back to 6–7 (with formic acid), and a 50 µl sample of 0.1 M EDTA·NaOH (pH 7.0, 25°C) has been added.

HPLC PURIFICATION OF InsP$_5$s

There are six *myo*-inositol pentakisphosphates, two of which possess a plane of symmetry; the remaining four form two pairs of enantiomers (see above). Consequently, non-chiral chromatographic techniques would be expected to resolve a maximum of four species of InsP_5: Ins(1,2,3,4,6)P_5, Ins(1,3,4,5,6)P_5 (*meso*-compounds which can, in theory at least, be purified completely by standard chromatographic techniques) and D- and L-Ins(1,2,3,4,5)P_5 and D- and L-Ins(1,2,4,5,6)P_5 (i.e., the best that non-chiral chromatography could achieve with these isomers is, for example, to obtain an Ins(1,2,4,5,6)P_5 peak that did not contain any of the other four non-enantiomeric InsP_5s, but could itself contain either D- or L-Ins(1,2,4,5,6)P_5). Several published methods have resolved four species of InsP_5. These include: (i) moving-paper iontophoresis (23); (ii) ion exchange HPLC utilizing an ion exchange column and nitric acid as an eluant [this requires a nitric acid-compatible HPLC system (24)]; (iii) anion exchange chromatography on 1 m long Dowex resin columns (the resin is in the chloride form and is eluted with HCl; run times extend to many hours (25)]; (iv) on commercial HPLC columns with HCl as an eluant [this requires an HCl-compatible solvent delivery system etc. (11)]. All of the HCl- or HNO_3-based systems require considerable care to avoid acid-catalysed migration of the phosphate groups in the Ins$P$s being purified. All of these systems have the significant advantage of resolving the InsP_5s in one run. We have developed an alternative set of conditions that can fully resolve all four species of InsP_5 by using the standard phosphate buffers and anion exchange HPLC columns that are commonly employed to separate InsP_4s and InsP_3s. The disadvantage of the procedure based on phosphate-buffered eluants is that two chromatographic steps are required; however, the resolution of the four InsP_5 species is complete even with well-used columns and no phosphate migration problems occur. On a standard SAX HPLC column (e.g., Partisil 10 SAX

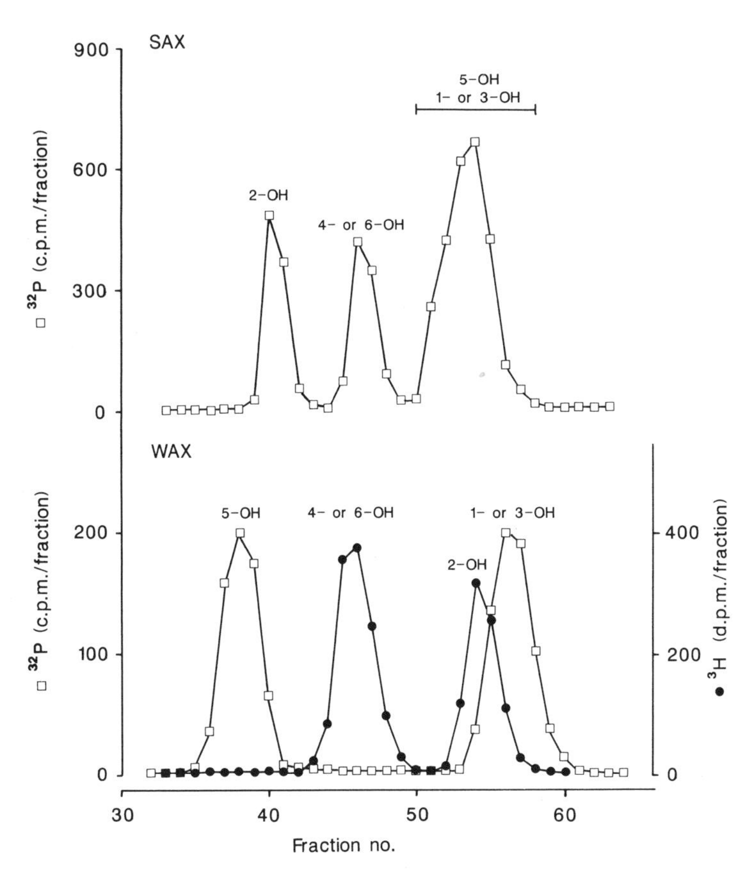

FIG. 1. Separation of inositol pentakisphosphates by HPLC. A mixture of ^{32}P-labelled Ins(1,2,3,4,6)P_5, D/L-Ins(1,2,4,5,6)P_5, D/L-Ins(1,2,3,4,5)P_5 and Ins(1,3,4,5,6)P_5 was applied to a Partisphere SAX HPLC column C (above) and eluted as described in the text. Fractions were collected every 0.5 min and counted for ^{32}P by Čerenkov counting. The peak that eluted last, which contained D/L-[^{32}P]Ins(1,2,4,5,6)P_5 and [^{32}P]Ins(1,2,3,4,6)P_5 was pooled, desalted, mixed with [^{3}H]Ins(1,3,4,5,6)P_5 and D]L-[^{3}H]Ins(1,2,4,5,6)P_5 and applied to a Partisphere WAX HPLC column (below). The column was eluted as described in the text and the fractions were counted for ^{3}H and ^{32}P by dual-label scintillation counting.

or Partisphere 5 SAX), InsP_5s can be effectively separated by eluting with a gradient of 0.6–1.0 M $(NH_4)_2HPO_4 \cdot H_3PO_4$ (pH 3.8, 22°C). The order of elution is Ins(1,3,4,5,6)P_5, D/L-Ins(1,2,3,4,5)P_5 and finally Ins(1,2,3,4,6)P_5 and D/L-Ins(1,2,4,5,6)P_5, which elute almost precisely together (see Fig. 1). On a standard WAX HPLC column (e.g., Partisphere WAX) eluted with a gradient of 0.3–0.4 M $(NH_4)_2$ $HPO_4 \cdot H_3PO_4$ (pH 3.2 at 22°C; these ionic strengths are appropriate for a new column, but the concentration of phosphate buffer will need to be reduced to maintain resolution as the column ages). The order of elution is Ins(1,2,3,4,6)P_5, D/L-Ins(1,2,3,4,5)P_5, and finally Ins(1,3,4,5,6)P_5 and D/L-Ins(1,2,4,5,6)P_5 eluting very close together. It is possible substantially to separate these last two isomers, particularly if a new 25 cm long column is being used, but this separation is never able to deal with samples containing greatly differing amounts of Ins(1,3,4,5,6)P_5 and D/L Ins(1,2,4,5,6)P_5, nor does it remain satisfactory as the column ages. Consequently, we routinely apply unknown samples (mixed with appropriate standards, see above) to a SAX HPLC column. If any material elutes in the Ins(1,2,3,4,6)/(1,2,4,5,6)P_5 region then a portion of the same sample is applied to a WAX column. Alternatively the sample applied to the SAX column could be desalted – from the scintillation fluid in which it was counted if necessary (10) – and reapplied to a WAX column that will now separate Ins(1,2,3,4,6)P_5 and D/L-Ins(1,2,4,5,6)P_5 to the two extremes of its InsP_5 range.

STEREO-SELECTIVE ASSIGNMENT OF INOSITOL PHOSPHATES

Enzymes are almost universally stereo-selective (although inositol monophosphate phosphatase is less so than most), and this means that it is insufficient to define an inositol phosphate as being D- or L-InsP_n. As far as a cell is concerned these are entirely different molecules; for example, L-Ins(1,4,5)P_3 (i.e., Ins(3,5,6)P_3) is a completely ineffective calcium mobilizer, in contrast to its enantiomer D-Ins(1,4,5)P_3 (26).

In theory, the handedness of a pure inositol phosphate (which is classified by reference to known "standard" compounds) can be established by a number of means. Ballou and co-workers (27) ultimately established the chirality of the PtdIns(4,5)P_2 head group by measuring the direction in which the plane of light from a plane-polarized source was rotated when it passed through a solution of iditol derived from PtdIns(4,5)P_2 (actually a fully acetylated derivative of the iditol was used); they compared this with standard independently prepared "L" and "D"-iditols. Note that this whole exercise could have been carried out with the unmodified Ins(1,4,5)P_3 head group, but a final classification would have required an independent synthesis of the D and L species of Ins(1,4,5)P_3, something that has only recently been achieved [see e.g., (26)].

The stereochemical classification of several InsP_5s has been done by measuring their specific optical rotations relative to reference compounds (18). A major draw back of this strategy is that sufficient pure material to produce several millilitres of a 1 mM solution is required. The stereo-selective assignment of D- and L-Ins(1)P has been achieved by use of a gas–liquid chromatography system utilizing a chiral stationary phase (28). This approach has not proved to be very versatile, probably because the methods of detection that have been developed for use with gas–liquid chromatography systems are not compatible with ^{3}H-labelled samples and they require micromolar quantities of appropriately derivatized material.

NMR techniques, when coupled with chirally selective shifting reagents, can also distinguish between enantiomers. However, both the compound of interest and each of the relevant "standards" must be available in milligram quantities and in a pure state. Moreover, the interactions of these molecules with the chiral-shifting reagent have to be fully characterized.

Finally, enzymes or binding proteins that are stereo-selective for either the original inositol phosphate or some derivative of them can be used. Again, a source of the relevant enantiomers in a resolved (although not necessarily pure) form is required, but the amounts needed can be very high and are determined by (i) the detection systems used to quantify the binding or formation of product, and (ii) the affinity of the proteins for their substrates or potential ligands (less than picomolar, under the best circumstances). To characterize [^{3}H]Ins*P*s extracted from [^{3}H]Ins-labelled cells, the most readily applied techniques employ stereo-selective enzymes or binding proteins.

Analysis of the structure of trace quantities of [^{3}H]Ins*P*s is usually achieved by identifying the ^{3}H-labelled polyols produced by a process of oxidation with periodate, reduction and dephosphorylation of the unknown inositol phosphate. This strategy was originally devised by Ballou et al. (27,29); because they worked with gram-scale quantities of material they established the chirality of the final products of this process by measuring their specific optical rotations (see above). The polyols generated by this process for all *myo*-inositol bis-, tris- and tetrakisphosphates are presented in Tables 1, 2 and 3. It should be emphasized that these are all starting from

TABLE 1. *The structures, and the polyols that would be derived from them, of all possible non-cyclic* myo-*inositol tetrakisphosphates*

Original inositol tetrakisphosphate	Enantiomer	Polyol derived from original InsP_4
Ins(1,2,3,4)P_4	Ins(1,2,3,6)P_4	L-Altritol
Ins(1,2,3,5)P_4	*meso*	Ins
Ins(1,2,3,6)P_4	Ins(1,2,3,4)P_4	D-Altritol
Ins(1,2,4,5)P_4	Ins(2,3,5,6)P_4	Ins
Ins(1,2,4,6)P_4	Ins(2,3,4,6)P_4	Ins
Ins(1,2,5,6)P_4	Ins(2,3,4,5)P_4	D-Glucitol[a]
Ins(1,3,4,5)P_4	Ins(1,3,5,6)P_4	Ins
Ins(1,3,4,6)P_4	*meso*	Ins
Ins(1,3,5,6)P_4	Ins(1,3,4,5)P_4	Ins
Ins(1,4,5,6)P_4	Ins(3,4,5,6)P_4	D-Iditol
Ins(2,3,4,5)P_4	Ins(1,2,5,6)P_4	L-Glucitol
Ins(2,3,4,6)P_4	Ins(1,2,4,6)P_4	Ins
Ins(2,3,5,6)P_4	Ins(1,2,4,5)P_4	Ins
Ins(2,4,5,6)P_4	*meso*	Ins
Ins(3,4,5,6)P_4	Ins(1,4,5,6)P_4	L-Iditol

[a] Sorbitol.

TABLE 2. *The structures, and the polyols that would be produced from them, of all possible non-cyclic isomers of* myo-*inositol trisphosphate*

Inositol trisphosphate	Enantiomer of that InsP_3	Polyol derived from that InsP_3
Ins(1,2,3)P_3	*meso*	Adonitol[a]
Ins(1,2,4)P_3	Ins(2,3,6)P_3	L-Altritol
Ins(1,2,5)P_3	Ins(2,3,5)P_3	D-Glucitol
Ins(1,2,6)P_3	Ins(2,3,4)P_3	D-Arabitol
Ins(1,3,4)P_3	Ins(1,3,6)P_3	L-Altritol
Ins(1,3,5)P_3	*meso*	Ins[a]
Ins(1,3,6)P_3	Ins(1,3,4)P_3	D-Altritol
Ins(1,4,5)P_3	Ins(3,5,6)P_3	D-Iditol
Ins(1,4,6)P_3	Ins(3,4,6)P_3	D-Iditol
Ins(1,5,6)P_3	Ins(3,4,5)P_3	Xylitol[a]
Ins(2,3,4)P_3	Ins(1,2,6)P_3	L-Arabitol
Ins(2,3,5)P_3	Ins(1,2,5)P_3	L-Glucitol
Ins(2,3,6)P_3	Ins(1,2,4)P_3	D-Altritol
Ins(2,4,5)P_3	Ins(2,5,6)P_3	L-Glucitol
Ins(2,4,6)P_3	*meso*	Ins[a]
Ins(2,5,6)P_3	Ins(2,4,5)P_3	D-Glucitol
Ins(3,4,5)P_3	Ins(1,5,6)P_3	Xylitol[a]
Ins(3,4,6)P_3	Ins(1,4,6)P_3	L-Iditol
Ins(3,5,6)P_3	Ins(1,4,5)P_3	L-Iditol[b]
Ins(4,5,6)P_3	*meso*	Xylitol[a,c]

[a] Optically inactive polyol.
[b] There is an error in (36).
[c] If Ins(4,5,6)P_3 is labelled with *myo*-[2-^{3}H]inositol, then upon oxidation with periodate the ^{3}H is lost.

myo-inositol; similar products could be obtained from other inositols with different phosphate arrangements. When this strategy was originally adapted to deal with trace quantities of ^{3}H-labelled inositol phosphates (30) the ability to distinguish between stereoisomers was lost (and enzymes specific for the original inositol phosphates were not available). The recent use of a stereo-selective yeast-derived polyol dehydrogenase (7,31,32) has regained the resolving power that this approach originally possessed, and therefore retains the great advantage of converting a wide variety of inositol phosphates into a relatively small family of polyols, many of which can be oxidized by a single dehydrogenase preparation with total D versus L selectivity [see Table 4; (33,34)]. Obviously, this approach is more adaptable than one in which enzymes that stereo-selectively metabolize a particular inositol phosphate are individually characterized and exploited as stereo-selective tools. (It is also less prone to circular pieces of logic: the process of establishing the specificity of the enzyme is usually closely linked to the characterization of the structure(s) of its substrate(s).)

The principle of the polyol dehydrogenase assay is simple: molecules recognized as substrates are oxidized to ketones in a β-NAD^+-linked reaction. The progress of the reactions can be easily monitored in a spectrophotometer

TABLE 3. *The structures, and the polyols that would be derived from them, of all possible non-cyclic myo-inositol bisphosphate*

Inositol bisphosphate	Enantiomer	Polyol derived from that InsP_2
Ins(1,2)P_2	Ins(2,3)P_2	Erythritol
Ins(1,3)P_2	*meso*	Adonitol[a]
Ins(1,4)P_2	Ins(3,6)P_2	D-Iditol + L-altritol[b]
Ins(1,5)P_2	Ins(3,5)P_2	Xylitol
Ins(1,6)P_2	Ins(3,4)P_2	Threitol
Ins(2,3)P_2	Ins(1,2)P_2	Erythritol
Ins(2,4)P_2	Ins(2,6)P_2	L-Arabitol
Ins(2,5)P_2	*meso*	D/L Glucitol glycerol[b]
Ins(2,6)P_2	Ins(2,4)P_2	D-Arabitol
Ins(3,4)P_2	Ins(1,6)P_2	Threitol
Ins(3,5)P_2	Ins(1,5)P_2	Xylitol
Ins(3,6)P_2	Ins(1,4)P_2	L-Iditol + D-altritol[b]
Ins(4,5)P_2	Ins(5,6)P_2	Threitol[c]
Ins(4,6)P_2	*meso*	Xylitol[c]
Ins(5,6)P_2	Ins(4,5)P_2	Threitol[c]

[a] Ribitol.
[b] Upon partial oxidation with periodate; complete oxidation yields volatile ^{3}H.
[c] If the inositol bisphosphates were labelled with ^{3}H in the C-2 position then these isomers will lose the ^{3}H during oxidation with periodate. The polyols would remain labelled if [^{3}H]InsP_2s contained the ^{3}H at the 6-C centre. All numbering is from the D-1 substitution site.

by measuring the accumulation of NADH or by detecting the formation of a ^{3}H-labelled ketone. A commercially available preparation of a yeast-derived L-iditol dehydrogenase (Sigma) oxidizes substrates with a specificity that is consistent with the rules set out by McCorkindale and Edson (35). The enzyme (known as L-iditol dehydrogenase; EC 1.1.1.14) requires a polyol (usually C_4–C_6) with a primary carbinol group at C-1 (i.e., adjacent to the site of oxidation) and a secondary carbinol group in the L-configuration (relative to C-1) at C-2 (the site of oxidation) and C-4 (with the exception of L-arabitol; see Table 4). In this case the D and L configurations are all defined relative to D- and L-glyceraldehyde.

Oxidation of Iditols

The commercially available, yeast-derived L-iditol dehydrogenase (L-iditol: NAD 2-oxidoreductase, also found in mammalian tissues (34,35)) is, as its name implies, distinguishable from a number of other polyol dehydrogenases by its ability to oxidize L- but not D-iditol. In a buffer containing 100 mM Tris·HCl (pH 8.3, 22°C), 20 mM β-NAD$^+$, 1.5 units L-iditol dehydrogenase/ml (units defined by Sigma), L-iditol (prepared from L-idose as described (7)) is oxidized with a first-order rate constant of 6.1%/min per unit (K_m approximately 30 mM, V_{max} 360 nmol/min per unit). Because the reaction

TABLE 4. *The groups of non-cyclic myo-inositol phosphates which can, upon oxidation with periodate, reduction and dephosphorylation, yield polyols that can be enantiomerically resolved by a yeast-derived polyol dehydrogenase; also given are the oxidation products formed by the enzyme*

Inositol phosphates yielding a given polyol	Polyols[a]	Product of oxidation by yeast-derived polyol dehydrogenase
D-Ins(1,4)P_2[b] D-Ins(1,4,5)P_3 D-Ins(1,4,6)P_3 D-Ins(1,4,5,6)P_4	D-Iditol CH_2OH *CH_2OH	Not oxidized
L-Ins(1,4)P_2[b] L-Ins(1,4,5)P_3 L-Ins(1,4,6)P_3 L-Ins(1,4,5,6)P_4	L-Iditol *CH_2OH CH_2OH	L-Sorbose *CH_2OH =O CH_2OH
L-Ins(1,3,4)P_3 L-Ins(1,2,4)P_3 L-Ins(1,2,3,4)P_4	D-Altritol CH_2OH * CH_2OH	Not oxidized
D-Ins(1,3,4)P_3 D-Ins(1,2,4)P_3 D-Ins(1,2,3,4)P_4	L-Altritol CH_2OH * CH_2OH	L-Tagatose CH_2OH O * CH_2OH
L-Ins(1)P[b] D-Ins(2,4,5)P_3[c] L-Ins(1,2,5)P_3 L-Ins(1,2,5,6)P_4	L-Glucitol CH_2OH * CH_2OH	Not oxidized

TABLE 4. (*continued*)

Inositol phosphates yielding a given polyol	Polyols[a]	Product of oxidation by yeast-derived polyol dehydrogenase	
D-Ins(1)*P*[b] L-Ins(2,4,5)P_3[c] D-Ins(1,2,5)P_3 D-Ins(1,2,5,6)P_4	D-Glucitol CH_2OH * CH_2OH	D-Fructose CH_2OH =O * CH_2OH	
L-Ins(2,4)P_2 D-Ins(1,2,6)P_3	D-Arabitol CH_2OH * CH_2OH	Not oxidized	
D-Ins(2,4)P_2 L-Ins(1,2,6)P_3	L-Arabitol CH_2OH * CH_2OH	L-Ribulose CH_2OH =O * CH_2OH	L-Xylulose CH_2OH O= CH_2OH

[a] The location of the hydrogen (or tritium) attached to the C-2 of the original *myo*-inositol is shown by an asterisk for both the polyols and the oxidation products. Chemical structures are shown as their Fischer projections.

[b] When only partially oxidized by periodate.

[c] There is an error in (29); D-Ins(2,4,5)P_3 gives L-glucitol.

possesses a 1:1 stoichiometry between accumulated moles of NADH and of ketone (7), the extent of reaction can be accurately determined by measuring the absorbance of the reaction mixture at 340 nm. A convenient concentration of L-iditol with which to start the assays is 200 μM: in a 1 ml assay this represents an amount of polyol that can be readily chromatographed on an analytical-scale HPLC column but is sufficient to avoid significant non-specific losses. If 200 nmol of L-iditol is completely oxidized to ketone, it would yield sufficient NADH to increase $A_{1cm,\,340}$ by 1.24 and is well within the first-order range of the enzyme.

Typically we run an assay as follows: 980 μl of 0.1 M Tris·HCl (pH 8.3, 22°C), 20 mM β-NAD^+, 1.5 units yeast-derived L-iditol dehydrogenase/ml are allowed to equilibrate in two 1 ml quartz cuvettes at 22°C in a thermostatically controlled dual-beam spectrophotometer set to measure the difference in absorbance at 340 nm between the two cuvettes. After 10–20 min equilibration (during which the differences in absorbance should be stable and set to zero), the substrate mixture should be added to one cuvette (in 20 μl of H_2O) and H_2O to the reference chamber. The substrate mixture normally contains 1000 d.p.m. of D-[^{14}C]iditol (prepared as described in (36)), 200 μM L-iditol, and a [^{3}H]iditol of unknown isomerism. If a dual-beam spectrophotometer is not available, then the absorbance at 340 nm of two cuvettes containing the reagents present before the addition of the substrate mixture described above is measured. Initially, the basal absorbance at 340 nm will change significantly; however, this drift should stabilize after 15–20 min, at which point the substrate is added to one of the cuvettes and monitoring of the production of NADH in the experimental cuvette and the basal drift in the control is continued. After 90–100 min, the rate of change of absorbance at 340 nm should fall to that of the control samples, or, in the case of the dual-beam protocol described above, fall to zero. Once this has happened, the contents of the assay cuvette are transferred to a microfuge tube and boiled for 3 min. The heat-denatured protein is pelleted by centrifugation and the supernatant desalted by being mixed with 3 ml of MB3A mixed-bed, ion exchange resin (Sigma) and allowed to stand for 2–3 h. The resin is washed and the sample freeze-dried, resuspended in 10 μl of H_2O and applied to a cation exchange HPLC column (Pb^{2+} mode: see below) that will separate the reactants from products. The percentage of the total ^{3}H-labelled sample eluting after 16–17 min [for conditions see below or (4), (10) or (36)] can be assumed to be equal to the percentage of L-[^{3}H]iditol in the original sample (as long as none of the internal D-[^{14}C]iditol has been oxidized and the oxidation of the carrier L-iditol has gone to completion).

Oxidation of Altritols

The oxidation of L- but not D-altritol conforms to the McCorkindale–Edson rules ((35) and see above), which predict the substrate specificity of L-iditol dehydrogenase, although the first-order rate constant against L-altritol (0.69%/min per unit) is substantially lower than that against L-iditol (6.1/min per unit). The K_m of the preparation is still in the millimolar range, and it oxidizes D-altritol at $\frac{1}{132}$nd of the rate of L-altritol, under first-order reaction conditions. The reaction is summarized in Table 4; D- and L-altritol can be prepared as described elsewhere (36). The product is L-tagatose, which can be resolved from altritol and allulose (another potential product of the oxidation) on the cation exchange HPLC column described below. Assays are run in a manner identical with that for L-iditol oxidation (see above) except (i) 9.5 units L-iditol dehydrogenase/ml is used and (ii) the assays are run for about 2.5 h. As a consequence of these alterations, the value of the ratio nmol NADH : nmol tagatose accumlated is not 1. The value of the ratio steadily decreases as the assay progresses and this is a consequence of the higher rate of NADH destruction by contaminating enzymes in the dehydrogenase preparation and the increased significance of the photochemical degradation of NADH during the slower progress of this oxidation. To calibrate the extent of the reaction accurately we routinely include 500–1000 d.p.m. of L-[^{14}C]altritol in the assay, along with 200 μM L-altritol and the ^{3}H-altritol of unknown isomerism. The percentage of the L-[^{14}C]altritol oxidized is then used to define the extent of the reaction. Under the conditions defined above, 86–92% of the initial L-[^{14}C]altritol is oxidized. If D-[^{3}H]altritol is included in the assays [see (36)] then less than 5% of this enantiomer is oxidized. If the concentration of enzyme is raised significantly above 9.5 units/ml, in an attempt to take the reaction to completion, then the proportion of D-altritol oxidized creates problems.

A commercially available, sheep-liver-derived preparation of L-iditol dehydrogenase (Sigma) will oxidize L- but not D-iditol with a first-order rate constant similar to that of the yeast-derived enzyme, ~12%/min per unit at 22°C (units defined by Sigma). However, it will oxidize L-altritol at only $\frac{1}{300}$th of this rate. Whether this is a consequence of a species-related difference in substrate specificity of a single enzyme or of more than one activity in the yeast-derived preparation is not yet clear.

It should be noted that D-Ins(1,3,4)P_3, D-Ins(1,2,4)P_3 and D-Ins(1,2,3,4)P_4 yield L-altritol upon oxidation with periodate, reduction and dephosphorylation.

Oxidation of Glucitol

In accordance with the substrate specificity rules defined above, D-glucitol (sorbitol) but not L-glucitol is oxidized by L-iditol dehydrogenase, yielding D-fructose and NADH (L. Stephens, unpublished data). Under the assay conditions defined above, the K_m of L-iditol dehydrogenase for D-glucitol is 25 mM and its V_{max} is 330 nmol/min per unit. The oxidation of L-glucitol is undetectable. D-[^{14}C]glucitol (available from Amersham) is used as an internal calibration mechanism for each assay, and the product of its oxidation, D-[^{14}C]fructose, is separated from glucitol by the cation-exchange HPLC system described below. The assay can be safely run to completion, without significant oxidation of L-glucitol, by incubating with 3 units polyol dehydrogenase/ml for 2 h.

Again it should be noted that D-Ins(2,4,5)P_3 (as well as L-Ins(1,2,5)P_3 and L-Ins(1,2,5,6)P_4) yield L-glucitol upon oxidation with periodate, reduction and dephosphorylation. [There is an assignment error over this point in one of the original papers (29).]

Oxidation of Arabitol

The oxidation of L-arabitol (but not D-) by L-iditol dehydrogenase is the only exception to the McCorkindale–Edson rules ((35); see p. 21) breaking the requirement for the C-4 hydroxyl moiety to be L with respect to C-1. This is as might be expected, because the two carbinols adjacent to the two primary hydroxyl groups in L-arabitol are both L with respect to their neighbouring terminals and the C-4s (counting from either end of the molecule), and in each case D with respect to their C-1 terminals. L-Arabitol can be oxidized at either "end" by L-iditol dehydrogenase, yielding either L-xylulose or L-ribulose (see Table 4). The apparent K_m and V_{max} for the oxidation of L-arabitol by the yeast-derived L-iditol dehydrogenase are 25 mM and 111 nmol/min per unit, respectively (under the assay conditions defined above). However, the overall kinetics of the oxidation of L-arabitol by L-iditol dehydrogenase are clearly complex, and appear to result from the process being an equilibrium, such that under the conditions of the assay (9.5 units L-iditol dehydrogenase/ml, 100 mM Tris·HCl (pH 8.3, 22°C), 20 mM β-NAD^+), after 2.5 h roughly only half of the total substrate was converted to product, despite the fact that NADH had ceased to accumulate (the enzyme had only inactivated by 7% in the 2.5 h). This equilibrium is shown directly by the observations that: (i) if some of the residual L-arabitol was repurified and added back to a subsequent identical assay then again roughly half was oxidized; and (ii) if either of the products was purified from the assay to detect L-arabitol oxidation and then added back, then L-arabitol accumulated during the course of a 2 h assay. It should be noted that if the L-[^{3}H]arabitol was derived from *myo*-[2-^{3}H]inositol-labelled inositol phosphates then the ^{3}H will be lost from the polyol during the formation of L-xylulose. Despite these problems the complete stereo-selectivity of the dehydrogenase preparation enables the chirality of unknown [^{3}H]arabitols to be established.

CHROMATOGRAPHY OF [^{3}H]POLYOLS AND [^{3}H]KETONES

Cation exchange HPLC columns in either the Ca^{2+} or Pb^{2+} forms can be used to separate carbohydrates. We invariably use the Pb^{2+} form, as these columns offer higher resolution separations of the compounds in which we are interested (4,7,36,37). The columns are eluted with H_2O at 0.2 ml/min; thermostatically regulated temperatures of about 25°C give the best results (L. Stephens, unpublished data). Samples are injected in 10 μl of H_2O and can be readily recovered after separation as the eluant is also H_2O. Elution times for most of the polyols mentioned have been reported. In addition (4,36,37), xylulose and fructose elute before inositol, whereas ribulose elutes after arabitol (but before glucitol; L. Stephens, unpublished data). The only unexpected problem of which we are aware that has affected the performance of these columns was caused by silver solder leaching from Waters (Millipore Corp. Milford, MA) HPLC pumps gradually causing recoveries of polyols through these columns to fall to zero (S. Shears, personal communication).

Acknowledgment L.R. Stephens is a Babraham Research Fellow.

REFERENCES

1. Irvine RF. The structure, metabolism and analysis of inositol lipids and inositol phosphates. In: Putney JW Jr, ed. *Phosphoinositides and receptor mechanisms*. New York: Alan R. Liss, 1986;89–107.
2. Shears SB, Parry JB, Tang EKY, Irvine RF, Michell RH, Kirk CJ. Metabolism of D-myo-inositol-1,3,4,5-tetrakisphosphate by rat liver homogenates including synthesis of a novel isomer of myo-inositol tetrakisphosphate. *Biochem J* 1987;246:139–147.
3. Balla T, Guillemette G, Baukal AJ, Catt KJ. Metabolism of inositol 1,3,4-trisphosphate to a new tetrakisphosphate isomer in angiotensin-stimulated adrenal glomerulosa cells. *J Biol Chem* 1987;262:9952–9955.
4. Stephens LR, Hawkins, PT, Barker CJ, Downes CP. Synthesis of *myo*-inositol 1,3,4,5,6-pentakisphosphate from inositol phosphates generated by receptor activation. *Biochem J* 1988;253:721–723.
5. Hansen CA, Dahl S, Huddell B, Williamson JR. Characterization of inositol 1,3,4-trisphosphate phosphorylation in rat liver. FEBS Lett 1987;236:53–56.
6. Shears SB. The pathway of *myo*-inositol 1,3,4-trisphosphate phosphorylation in liver. *J Biol Chem* 1989;264:19879–19886.
7. Stephens LR, Hawkins PT, Carter N, Chahwala SB, Morris AJ, Whetton AD, Downes CP. L-*myo*-inositol 1,4,5,6-tetrakisphosphate is present in both mammalian and avian cells. *Biochem J* 1988;249:271–282.
8. Stephens LR, Hawkins PT, Morris AJ, Downes CP. L-*myo*-inositol 1,4,5,6-tetrakisphosphate (3-hydroxyl)kinase. *Biochem J* 1988;249:282–292.
9. Rhu SO, Lee SY, Lee K, Rhee SG. Catalytic properties of inositol trisphosphate kinase: activation by Ca^{2+} and calmodulin. *Fed Am Soc Exp Biol J* 1987;1:388–393.
10. Stephens LR, Downes CP. Precursor–product relationships amongst inositol polyphosphates. *Biochem J* 1990;265:435–452.
11. Mayr GW. A novel metal–dye detection system permits picomolar-range hplc analysis of inositol polyphosphates from non-radioactively labelled cell or tissue specimens. *Biochem J* 1988;254:585–591.
12. Sharpes ES, McCarl RL. A high performance liquid chromatographic method to measure ^{32}P incorporation into phosphorylated metabolites in cultured cells. *Anal Biochem* 1982;124:421–424.
13. Palmer S, Hughes KT, Lee DY, Wakelam MJO. Development of a novel Ins(1,4,5)P_3-specific binding assay: its use to determine the intracellular concentration of Ins(1,4,5)P_3 in unstimulated and vasopressin-stimulated rat hepatocytes. *Cellular Signalling* 1989;1: 147–156.
14. Wreggett KA, Howe LR, Moore JP, Irvine RF. Extraction and recovery of inositol phosphates from tissues. *Biochem J* 1987;245:933–934.
15. Irvine RF, Anggard EE, Letcher AJ, Downes CP. Metabolism of inositol 1,4,5-trisphosphate and inositol 1,3,4-trisphosphate in rat parotid glands. *Biochem J* 1985;229:505–511.
16. Dean NM, Moyer JD. Separation of multiple isomers of inositol phosphates formed in GH_3 cells. *Biochem J* 1987;242:361–366.
17. Johnson LF, Tate ME. The structure of 'phytic acids'. *Can J Chem* 1969;47:63–73.
18. Irving GCJ, Cosgrove DJ. Inositol phosphates phosphatases of microbiological origin: the inositol pentaphosphate products of *Aspergillus ficuum* phytases. *J Bacteriol* 1972;112:434–438.
19. Tomlinson RV, Ballou CE. *Myo*-inositol polyphosphate intermediates in the dephosphorylation of phytic acid by phytase. *Biochemistry* 1962;1:166–171.
20. Lim PE, Tate ME. The phytases II. *Biochim Biophys Acta* 1973;302:316–328.
21. Posternak S. Synthesis of inosite hexaphosphoric acid. *J Biol Chem* 1921;46:453–457.
22. Phillippy BQ, White KD, Johnston MR, Tao SH, Fox MRS. Preparation of inositol phosphates from sodium phytate by enzymatic and non-enzymatic hydrolysis. *Anal Biochem* 1987;162:115–121.

23. Tate ME. Separation of *myo*-inositol pentakisphosphates by moving paper electrophoresis. *Anal Biochem* 1968;23:141–149.
24. Phillippy BQ, Bland JM. Gradient ion chromatography of inositol phosphates. *Anal Biochem* 1988;175:162–166.
25. Cosgrove DJ. Ion-exchange chromatography of inositol polyphosphates. *Ann NY Acad Sci* 1969;165:677–686.
26. Taylor CW, Berridge MJ, Cooke AM, Potter BVL. Inositol 1,4,5-trisphosphorothioate, a stable analogue of inositol trisphosphate which mobilizes intracellular calcium. *Biochem J* 1989;254:645–650.
27. Grado C, Ballou CE. *Myo*-inositol phosphates obtained by alkaline hydrolysis of beef brain phospholipids. *J Biol Chem* 1961;236:54–60.
28. Loewus MW, Sasaki K, Leavitt AL, Munsell L, Sherman WR, Loewus FA. Enantiomeric form of *myo*-inositol-1-phosphate produced by *myo*-inositol phosphate synthase and *myo*-inositol kinase in higher plants. *Plant Physiol* 1982;70:1661–1663.
29. Tomlinson RV, Ballou CE. Complete characterization of *myo*-inositol polyphosphates from beef brain phosphoinositides. *J Biol Chem* 1961;236:1902–1906.
30. Irvine RF, Letcher AJ, Lander DJ, Downes CP. Inositol trisphosphates in carbachol-stimulated rat parotid glands. *Biochem J* 1984;223:237–243.
31. Barker CJ, Morris AJ, Kirk CJ, Michell RH. Inositol tetrakisphosphates in WRK-1-cells. *Biochem Soc Trans* 1988;16:984–985.
32. Balla T, Hunyady L, Baukal AJ, Catt KJ. Structures and metabolism of inositol tetrakisphosphates and inositol pentakisphosphates in bovine adrenal glomerulosa cells. *J Biol Chem* 1989;264:9386–9390.
33. Arcus AC, Edson NL. Polyol dehydrogenases. 2. The polyol dehydrogenase of *Acetobacter suboxydans* and *Candida utilis*. *Biochem J* 1956;64:385–394.
34. Smith MG. Polyol dehydrogenases. 4. Crystallization of the L-iditol dehydrogenase of sheep liver. *Biochem J* 1962;83:135–144.
35. McCorkindale J, Edson NL. Polyol dehydrogenases. 1. The specificity of rat liver polyol dehydrogenase. *Biochem J* 1954;57:518–523.
36. Stephens LR, Hawkins PT, Downes CP. An analysis of *myo*-[^{3}H]inositol trisphosphates found in *myo*-[^{3}H]inositol prelabelled avian erythocytes. *Biochem J* 1989;262:727–737.
37. Stephens LR, Hawkins PT, Downes CP. Metabolic and structural evidence for the existence of a third species of polyphosphoinositide in cells: D-phosphatidyl-*myo*-inositol 3-phosphate. *Biochem J* 1989;259:267–276.

Methods in Inositide Research,
edited by Robin F. Irvine.
Raven Press, Ltd., New York © 1990

3

Preparation of ^{32}P-Labelled Inositol 1,4,5-Trisphosphate and ^{14}C-Labelled Inositol 1,4-Bisphosphate

Andrew J. Letcher, Leonard R. Stephens, and Robin F. Irvine

Department of Biochemistry, Institute of Animal Physiology and Genetics Research, Babraham, Cambridge CB2 4AT, UK

The reason for this chapter is that for most routine HPLC analysis of inositol phosphates it is very useful to include an internal marker for inositol monophosphate (Ins*P*), bisphosphate (InsP_2) and trisphosphate (InsP_3). As the inositol phosphates being analysed are usually labelled with tritium, the internal markers are most useful if radiolabelled with ^{14}C or ^{32}P. ^{14}C-labelled Ins(3)*P* is commercially available, but we know of no source of [^{14}C]Ins(1,4)P_2, and so a simple preparation of this compound, starting with ^{14}C-labelled phosphatidylinositol (PtdIns) (commercially available) is described. ^{32}P-labelled Ins(1,4,5)P_3 is commercially available, and can also be prepared as described by Downes et al. (1), but using the modified, low chloride, medium of King et al. (2); this is a method which can be miniaturized for small-scale preparation of [^{32}P]Ins(1,4,5)P_3. The Ins(1,4,5)P_3 can be desalted, either as described below or by the method described in Chapter 2. But we have, over the last few years, routinely employed an alternative preparation that does not require the handling of large quantities of [^{32}P]P_i.

PREPARATION OF ^{32}P-LABELLED D-*MYO*-INOSITOL 1,4,5-TRISPHOSPHATE

^{32}P-labelled D-*myo*-inositol 1,4,5-trisphosphate ([^{32}P]Ins(1,4,5)P_3) is prepared using rat brain phosphatidyl inositol 4-phosphate 5-kinase (EC 2.7.1.68) to catalyse the phosphorylation of bovine brain phosphatidyl inositol 4-monophosphate (PtdIns(4)*P*), by ^{32}P-labelled ATP, to form (5-^{32}P)-labelled phosphatidyl inositol 4,5-trisphosphate ([5-^{32}P]PtdIns(4,5)P_2). This is subsequently deacylated, and then the glycerol moiety is removed by mild oxidation with periodate.

PtdIns(4)*P* is prepared from a bovine brain inositide fraction (3) by column chromatography on immobilized neomycin [(4); and see Chapter 1 for further details].

Under the conditions used, the presence of phosphatidylethanolamine (PtdEtn) in a 10:1 excess over PtdIns(4)*P* in the incubation mixture greatly enhances the rate of incorporation of [^{32}P]P_i into [^{32}P]PtdIns(4,5)P_2 over that achieved with PtdIns(4)*P* alone. PtdEtn is obtained from porcine liver by column chromatography on first alumina (5) and then silicic acid (6). PtdEtn from several other sources has been used but all resulted in lower rates of incorporation of [^{32}P]P_i into [^{32}P]PtdIns(4,5)P_2, when compared with PtdEtn from pig liver.

Rat brain PtdIns*P* kinase is prepared by purification from a whole rat brain (approximately 2 g) using a high-performance liquid chromatography (HPLC) ion exchange column Bio-Gel TSK-DEAE-5 PW (BioRad Laboratories, Richmond, CA). This preparation is described in Chapter 16. We have also used the fraction eluted from a soft DEAE-cellulose gel (see Chapter 16) successfully and, although this is less active, it dispenses with the need for expensive HPLC columns.

Preparation of [^{32}P]PtdIns(4,5)P_2

PtdIns(4)*P* (200 nmol) and PtdEtn (2000 nmol) are pooled as solutions in chloroform in a thin-walled glass tube and dried at room temperature in a current of nitrogen gas. H_2O (0.5 ml) is added and the tube is placed in a bath sonicator at room temperature for 2 min, or until all the lipid has been displaced from the tube wall. The sample is stored on ice, and to it are added the other components of the incubation mixture (see below) giving a final volume of 1.3 ml.

The incubation mixture is 40 mM Tris-acetate (pH 7.4), 15 mM magnesium acetate, 40 μM ATP, 1.85 MBq (50 μCi) adenosine 5′-[γ-^{32}P]triphosphate, triethylammonium salt in aqueous solution (Amersham International, p.l.c.), and 0.5 ml of an enzyme preparation from rat brain. The incubation is started by adding the enzyme and is continued for 4 h at room temperature.

The reaction is stopped by adding 5 ml of chloroform/methanol (1:2, v/v), 20 μl of 12 M HCl. After vortexing, the tube is stored on ice for 10 min, after which 1.67 ml of chloroform and then 1.67 ml of 1 M HCl are added to complete the extraction mixture and to form two phases. The lower phase is washed six times with theoretical upper phase chloroform/methanol/1 M HCl (3:48:47, by vol.), with mixing and centrifugation after each wash. The final upper phase is removed by aspiration and the lower phase dried thoroughly to remove all the HCl. The [^{32}P]PtdIns(4,5)P_2 is dissolved in a small amount of chloroform.

Preparation of ^{32}P-Labelled Glycerophosphatidylinositol 4,5-Bisphosphate ([5-^{32}P]GroPIns(4,5)P_2)

The [^{32}P]PtdIns(4,5)P_2 is transferred to a vessel suitable for monomethylamine treatment, and the chloroform is evaporated. Monomethylamine reagent (3 ml) [prepared as in (7)] is added, and the vessel is securely stoppered and incubated at 53°C for 60 min in a water-bath. The vessel is cooled on ice and the sample is dried using a rotary evaporator, starting with the vessel immersed in cold H_2O and finally heating at 50°C to remove all the monomethylamine. The procedure of gradually heating the flask prevents "bumping" and subsequent loss of the sample. Alternatively, the monomethylamine reagent may be removed by a Vac-Fuge.

The [5-^{32}P]GroPIns(4,5)P_2 sample thus obtained is dissolved in 2 ml of water and 2.4 ml of *n*-butanol/petroleum ether (40–60°C)/ethyl formate (20:4:1, by vol.) is added (7). The mixture is thoroughly shaken and is allowed to stand to form two clear phases. The upper phase containing *n*-methyl fatty acid amides is discarded and the lower phase is washed a second time with 1.5 ml of the same *n*-butanol/petroleum ether (40–60°C)/ethyl formate mixture before being taken to dryness.

Preparation of [^{32}P]Ins(1,4,5)P_3

The glycerol moiety is removed from the [5-^{32}P]GroPIns(4,5)P_2 by a modification of the mild periodate oxidation method of Brown and Stewart (8). To the dry [5-^{32}P]GroPIns(4,5)P_2 is added 1 ml of 50 mM sodium periodate. After mixing to dissolve the [5-^{32}P]GroPIns(4,5)P_2, the tube is placed in the dark for 30 min at room temperature. If oxidation is allowed to continue for longer the periodate will start to oxidize the inositol ring. Ethylene glycol (150 μl of 10% (v/v)) is added to stop the reaction and, after 15 min, 0.5 ml of 1% 1,1-dimethylhydrazine/formic acid (pH 4) (8)) is added, and the mixture is allowed to stand for a further 60 min.

The mixture is passed through 30 ml of a cation exchange resin (BioRad AG 50 W-×8 200–400, in the acid form) in a 1.6 cm diameter glass column. (Alternatively, the Dowex can be added directly to the sample and filtered off.) The column is washed with 30 ml of H_2O, the pooled eluates being adjusted to pH 6 with dilute potassium hydroxide and evaporated to dryness on a rotary evaporator.

The [^{32}P]Ins(1,4,5)P_3 is now dissolved in a small volume of H_2O for application to HPLC. Typically, 2 ml of H_2O is used and the purification of the sample is carried out in two runs. The column used is a 25 cm × 4.6

mm Partisil 10 SAX HPLC column supplied by Technicol Ltd, Higher Hillgate, Stockport, Cheshire, UK. A flow-rate of 1.5 ml/min is used and the column is equilibrated with 0.24 M sodium dihydrogenorthophosphate (adjusted to pH 3.8 with orthophosphoric acid). The 1 ml sample is injected on to the column and the eluting buffer is immediately increased to 0.55 M sodium dihydrogenorthophosphate (pH 3.8, adjusted as before). An isocratic run in this buffer is continued for 20 min. A 20 μg standard of cold ATP is included in the sample injection and this is located with an ultraviolet detector at 254 nm, and would normally be eluted 10 min after injection. Fractions are collected at 15 s intervals from the appearance of the ATP peak, and collection is carried out for 10 min.

The fractions are collected in 6 ml plastic vials and are counted by Čerenkov radiation; the [^{32}P]Ins(1,4,5)P_3 usually elutes at fractions 20–25. The relevant fractions are pooled, diluted 10 times with H_2O and loaded on to an Amprep strong anion exchange (SAX) 100 mg cartridge (Amersham International, p.l.c.) pre-washed with 3 M ammonium formate, 0.1 M formic acid followed by 10 ml H_2O.

The loaded Amprep is washed with 20 ml of 0.15 M ammonium formate, 0.015 M formic acid to remove the phosphate, and the [^{32}P]Ins(1,4,5)P_3 is eluted with 10 ml of 1 M triethylamine bicarbonate (freshly prepared by bubbling CO_2 through a 14% (w/v) triethylamine solution). The triethylamine bicarbonate is removed using a rotary evaporator and the residual [^{32}P]Ins(1,4,5)P_3 is dissolved in a small volume of water. A typical yield is 0.1 MBq [^{32}P]Ins(1,4,5)P_3 starting with 1.85 MBq of [^{32}P]ATP; i.e. 5.4% incorporation. The [^{32}P]Ins(1,4,5)P_3 can be converted to [^{32}P]Ins(1,3,4,5)P_4 if required by the method of Irvine et al. (9).

PREPARATION OF ^{14}C-LABELLED D-*MYO*-INOSITOL 1,4-BISPHOSPHATE ([^{14}C]INS(1,4)P_2)

[^{14}C]Ins(1,4)P_2 is prepared by using human erythrocyte phosphatidylinositol kinase (PtdIns 4-kinase, EC 2.7.1.67) to catalyse the phosphorylation of ^{14}C-labelled phosphatidyl inositol to [^{14}C]PtdIns(4)P, and then subsequent deacylation and glycerol removal by oxidation with mild periodate.

Preparation of Human Erythrocyte "Packed Ghosts"

Fresh human blood (20 ml) is added to 3 ml of acid/dextrose (2.5% (w/v) trisodium citrate, 1.5% (w/v) citric acid, 2% (w/v) glucose in H_2O). The blood is centrifuged at 500 *g* for 20 min. The supernatant and white cell "coat" are removed by aspiration and the pellet is washed twice with 10 mM Hepes, 130 mM potassium chloride, 10 mM sodium chloride (pH 7.0) at 4°C. The loose pellet is then poured into 100 ml of 10 mM Tris·HCl, 1 mM EDTA (pH 7) to lyse the erythrocytes into ghosts.

The ghosts are centrifuged at 100,000 *g* for 20 min and washed three times with the Tris/EDTA solution. The final pellet is vortexed and stored frozen as "packed ghosts."

Preparation of [^{14}C]PtdIns(4)*P*

PtdIns (74 kBq (2 μCi)) radiolabelled with ^{14}C in the inositol ring, specific activity > 8.1 GBq/mmol, is dried down and resuspended in 50 μl of 4.8% (w/v) Triton X-100.

The incubation is carried out in 400 μl, giving final concentrations of 15 mM 2-mercaptoethanol, 0.1 mM phenylmethylsulphonylfluoride (PMSF), 1 mg bovine serum albumin/ml, 5 mM ATP, 10 mM creatine phosphate, 5 units creatine phosphokinase/ml, 80 mM potassium chloride, 6 mM magnesium chloride, 50 mM Hepes (pH 7.0), 2 mM EGTA. This mixture is made up in 325 μl to which is added the [^{14}C]PtdIns in 4.8% Triton X-100 and 25 μl packed ghosts. Incubation is for 90 min at 37°C.

The reaction is stopped by adding 3.75 volumes (1.5 ml) chloroform/methanol/12 M HCl (200:400:1, by vol.) and 200 nmol of unlabelled PtdIns(4)*P* as a carrier. After the mixture has stood for 10 min, 0.5 ml of chloroform, and then 0.5 ml of 0.1 M HCl are added to form two phases. After being shaken, the tube is spun in a bench-top centrifuge, the upper phase and interface are removed by aspiration, and the lower phase is washed with (1 mM EDTA, 0.1 M HCl)/methanol (0.9:1, v/v).

The lower phase is evaporated to dryness and loaded on to a thin-layer chromatography (TLC) plate (Kieselgel $60F_{254}$ 20 cm × 20 cm obtainable from Merck, West Germany) that has been sprayed with 1% (w/v) potassium oxalate and heated at 100°C for 60 min. External standards of PtdIns and PtdIns*P* are also spotted on to the plate. The plate is developed in chloroform/methanol/4 M ammonium hydroxide (9:7:2, by vol.) for 60 min and is allowed to dry. The sample lane is covered with a glass plate while the PtdIns and PtdIns*P* standard lanes are sprayed with a phosphate-detecting spray (10).

The area of the sample lane corresponding to PtdIns*P* is scraped off and transferred to a suitable vessel for deacylation.

Preparation of [^{14}C]Gro*P*Ins(4)*P*

Deacylation of the [^{14}C]PtdIns(4)*P* is carried out by adding 3 ml of monomethylamine reagent (7) directly to the TLC scrapings. It has been established that the presence of silicic acid affects neither the deacylation reaction nor the subsequent recovery of [^{14}C]Gro*P*Ins(4)*P* (our unpublished results).

The deacylation step is carried out for 60 min at 53°C, after which the monomethylamine reagent is removed by drying on a rotary evaporator, starting at room temperature and gradually raising the temperature to 50°C.

The residue is dissolved in 2 ml of H_2O and 2.4 ml of *n*-butanol/petroleum ether (40–60°C)/ethyl formate (20:4:1, by vol.) is added. After mixing, the two phases are allowed to separate. The upper phase is removed by aspiration and the lower phase is given a further wash with 1.5 ml of the *n*-butanol/petroleum ether/ethyl formate mixture. The final lower phase is then dried *in vacuo*.

Preparation of [^{14}C]Ins(1,4)P_2

The glycerol moiety is removed from the [^{14}C]Gro*P*Ins(4)*P* by a modification of the mild periodate oxidation method of Brown and Stewart (8) exactly as described above for [5-^{32}P]Gro*P*Ins(4,5)P_2.

After passing through acid Dowex and having been neutralized with potassium hydroxide, the [^{14}C]Ins(1,4)P_2 is dissolved in 2 ml of H_2O ready for HPLC. The column used, as before, is a 25 cm × 4.6 mm Partisil 10 SAX column (Technicol Ltd). The flow-rate is 1.5 ml/min and the column is equilibrated with 0.1 M sodium dihydrogenorthophosphate adjusted to pH 3.8 with orthophosphoric acid. After loading, the eluting buffer is increased to 0.24 M sodium dihydrogenorthophophate (pH 3.8) and a 15 min isocratic run is carried out. A parallel run is carried out with an ADP standard (20 mg). As Ins(1,4)P_2 elutes shortly after ADP, its expected elution time can be estimated and fractions are collected every 15 s in the sample run from 1 min before the demonstrated elution time of ADP. The ADP is located with an ultraviolet detector at 254 nm.

A portion (1%) of each fraction is added to scintillation fluid for counting. The relevant fractions are pooled, diluted 10-fold with H_2O and loaded on to an Amprep SAX 100 mg cartridge (Amersham International, p.l.c.). The phosphate is eluted with 0.15 M ammonium formate, 0.015 M formic acid, and the [^{14}C]Ins(1,4)P_2 is eluted in 10 ml of 1 M triethylamine bicarbonate (freshly prepared by bubbling CO_2 through 14% (v/v) triethylamine). A typical yield is 15–18 kBq of [^{14}C]Ins(1,4)P_2 from 74 kBq of [^{14}C]PtdIns.

Finally, it should be noted that if the sole use of the [^{14}C]Ins(1,4)P_2 and [^{32}P]Ins(1,4,5)P_3 is as internal markers for HPLC, neither of them need be purified by HPLC as described; a single run of each with some [^{3}H]Ins(1,4)P_2 or [^{3}H]Ins(1,4,5)P_3 from commercial sources will establish whether the major radioactive compound is the desired one, and if it is, then each can be used in impure form as an internal marker.

Acknowledgments We thank Nullin Divecha for help and discussions; L.R.S. is a Babraham Fellow.

REFERENCES

1. Downes CP, Mussat MC, Michell RH. The inositol trisphosphate phosphomonoesterase of the human erythrocyte membrane. *Biochem J* 1982;203:169–177.
2. King CE, Stephens LR, Hawkins PT, Guy GR, Michell RH. Multiple pools of phosphoinositides and phosphatidate in human erythrocytes incubated in a medium that permits rapid transmembrane exchange of phosphate. *Biochem J* 1987;244:209–217.
3. Folch J. Complete fractionation of brain cephalin: isolation from it of phosphatidylserine, phosphatidylethanolamine and diphosphoinositide. *J Biol Chem* 1949;177:497–504.
4. Palmer FB. Chromatography of acidic phospholipids on immobilized neomycin. *J Lipid Res* 1981;22:1296–1300.
5. Irvine RF, Hemington N, Dawson RMC. The hydrolysis of phosphatidylinositol by lysosomal enzymes of rat liver and brain. *Biochem J* 1978;176:475–484.
6. Hanahan DJ, Dittmer JC, Warashina E. A column chromatographic separation of classes of phospholipids. *J Biol Chem* 1957;228:685–700.
7. Clarke NG, Dawson RMC. Alkaline $O \rightarrow N$-transacylation. A new method for the quantitative deacylation of phospholipids. *Biochem J* 1981;195:301–306.
8. Brown DM, Stewart JC. The structure of triphosphoinositide from rat brain. *Biochim Biophys Acta* 1966;125:413–421.
9. Irvine RF, Letcher AJ, Lander DJ, Berridge MJ. Specificity of inositol phosphate-stimulated Ca^{2+} mobilization from Swiss-mouse 3T3 cells. *Biochem J* 1986;240:301–304.
10. Vaskovsky VE, Kostetsky EY. Modified spray for the detection of phospholipids on thin-layer chromatograms. *J Lipid Res*. 1968;9:396.

Methods in Inositide Research,
edited by Robin F. Irvine.
Raven Press, Ltd., New York © 1990

4

Separation of Higher Inositol Phosphates by Polyethyleneimine-Cellulose Thin-Layer Chromatography and by Dowex Chloride Column Chromatography

Catherine E. L. Spencer, Leonard R. Stephens, and Robin F. Irvine

Department of Biochemistry, AFRC Institute of Animal Physiology and Genetics Research, Babraham, Cambridge CB2 4AT, UK

To date, the simplest and quickest method for analysing the inositol phosphate content of numerous samples has been anion exchange chromatography using small open-topped columns (1). However, on standard batch-eluted formate-form Dowex columns it is difficult to resolve inositol tetrakisphosphates (InsP_4), inositol pentakisphosphates (InsP_5) and inositol hexakisphosphate (InsP_6). We have therefore sought: (i) a more rapid method of separating inositol phosphates by thin-layer chromatography, which can be used for samples obtained from quenched tissues, but which we employ primarily for assaying phosphatases and kinases *in vitro;* and (ii) a very precise separation of inositol phosphates from trisphosphates to InsP_6 on Dowex columns which can also be used for quenched-tissue samples and for enzyme assays *in vitro*.

POLYETHYLENEIMINE-CELLULOSE CHROMATOGRAPHY

We have adapted a method using polyethyleneimine (PEI)-cellulose plates, previously used to separate inositol monophosphates (Ins$P$s), inositol bisphosphates (InsP_2s) and inositol trisphosphates (InsP_3s) (2) and Ins(1,3,4,5)P_4 from ATP (3). By developing the plates in HCl it is possible to separate Ins(1,4,5)P_3 from Ins(1,3,4,5)P_4 and InsP_5 from InsP_6, whilst they are still in the presence of conventional assay reagents. This has enabled us to develop improved assays for Ins(1,4,5)P_3 3-kinase and InsP_5 kinase.

Ins(1,4,5)P_3 3-Kinase Assays

Assays are carried out in 15 μl of reaction mixture containing 5 μl of enzyme or tissue supernatant, 50 mM Hepes/NaOH (pH 7.5, 25°C), 1 mM $MgCl_2$, 10 mM EGTA, 24 mM 2-mercaptoethanol, 1 mg bovine serum albumin/ml, 5 mM MgATP, 20 μM Ins(1,4,5)P_3 plus [^{3}H]Ins(1,4,5)P_3 (5000 d.p.m./assay tube). After incubation at 37°C for 20 min, the assays are stopped by adding 10 μl of 0.2 M HCl, 300 μM Ins(1,4,5)P_3, 300 μM Ins(1,3,4,5)P_4.

InsP_5 Kinase Assays

Assays are carried out in 15 μl of reaction mixture containing 5 μl of enzyme or tissue supernatant (usually from *Dictyostelium discoideum*), 3 mM ATP, 4 mM $MgCl_2$, 25 mM Hepes/NaOH (pH 7.0, 25°C), 1 mM EGTA, 80 mM KCl, 1 mM dithiothreitol, 1 mg bovine serum albumin/ml and 3000–6000 d.p.m. [^{3}H]InsP_5/assay tube. Reactions are quenched with 10 μl of 0.2 M HCl, 300 μM InsP_5, 300 μM InsP_6.

Thin-Layer Chromatography

The samples are dried down completely in standard 1.5 ml microfuge tubes using a Savant Speed-Vac concentrator, then resuspended in 5 μl of 0.3 M HCl, 0.2 M KH_2PO_4 (the inclusion of KPO_4 improves resolution by preventing smearing of the inositol phosphates on the plates). The samples are sonicated for a few seconds, centrifuged briefly in a microfuge, and then 2 μl portions are loaded, as single applications, at 8–9 mm intervals across the width of a PEI-cellulose-coated plastic plate.

The plates are developed in 0.5 M HCl (Ins(1,4,5)P_3 3-kinase assays) or 1.0 M HCl (InsP_5 kinase assays), dried in air at 60–80°C then sprayed to detect phosphate-containing compounds using the method described by Clarke and Dawson (4). The spray is a freshly prepared mixture of three parts H_2O/concentrated $HClO_4$/concentrated HCl (100:10:1, by vol.) to one part ammonium molybdate/$HClO_4$ (4%:9%, w/v, in H_2O). After drying in air at 60–80°C, the plates are exposed to ultraviolet light for 0.5–2 min, after which phosphate-containing compounds, including the inositol phosphate carriers, can be visualized directly as blue spots against a white background, with as little as 0.3 nmol of Ins(1,4,5)P_3 (Fig. 1).

The quantity of product formed or substrate remaining in each assay is determined by cutting out the appropriate regions in each lane. Inositol phosphates are extracted from the resulting segments by shaking them with 0.5 ml of concentrated HCl for 10–15 min, then 5 ml of H_2O and 6 ml of Packard Scintillant 299 are added and the samples counted for ^{3}H. Typically, more than 90% of the ^{3}H applied to the plates can be recovered. We originally

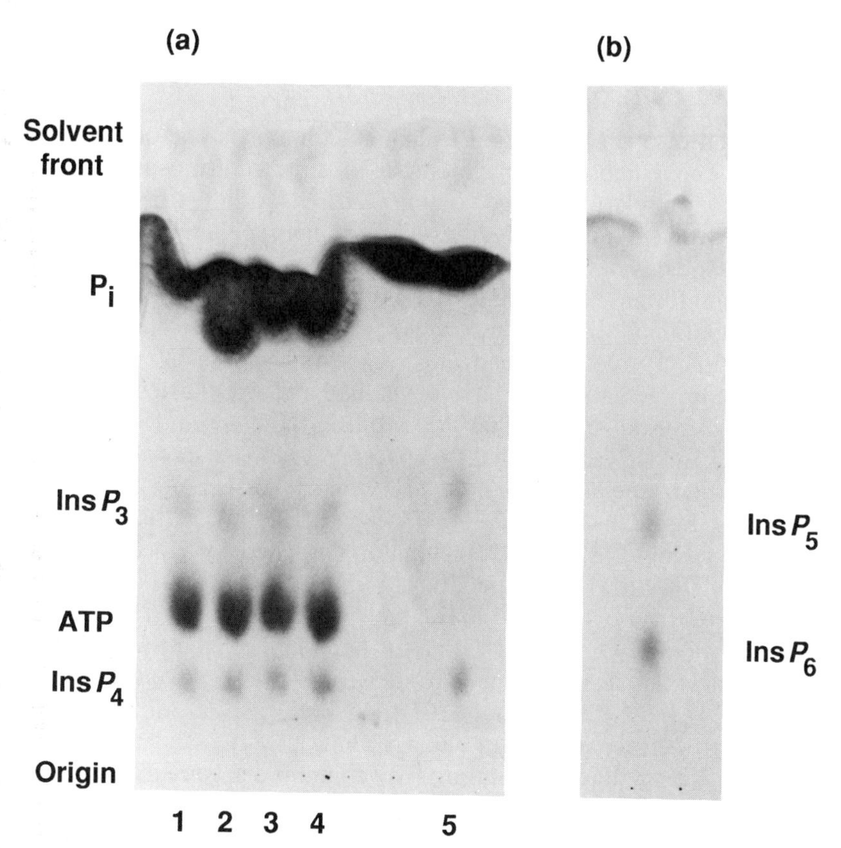

FIG. 1. **(a)** PEI-cellulose plate developed in 0.5 M HCl and sprayed for phosphate. Lanes 1–4, Ins(1,4,5)P_3 3-kinase assays; lane 5, HPLC-purified Ins(1,4,5)P_3 + HPLC-purified Ins(1,3,4,5)P_4. **(b)** PEI-cellulose plate developed in 1.0 M HCl and sprayed to identify phosphate. Mixture of HPLC-purified Ins(1,2,4,5,6)P_5 and InsP_6.

added ^{32}P-labelled inositol phosphates at the quenching step to calculate the recovery from each segment, but this was found not to be necessary due to the high recoveries obtained.

Origin of Marker Compounds

Commerically available inositol phosphates can be used as markers but phytic acid hydrolysate made by autoclaving a solution of InsP_6 can be crudely separated into its constituent families of inositol phosphates (i.e. InsP_2s, InsP_3s etc.) by Dowex column chromatography (see below), and the resulting mixtures of isomers yield clear results when used as marker compounds on the plates.

COLUMN CHROMATOGRAPHY OF HIGHER INOSITOL PHOSPHATES

Batch elution of open-topped ion exchange columns is one of the most reproducible, convenient and rapid methods for separating inositol phosphates when they are in relatively large volumes (200–300 μl) of either assay buffer or cellular extract. Traditionally, the anion exchange columns used to resolve inositol phosphates have been in the formate form and eluted with mixtures of ammonium formate and formic acid (5); although these systems have the advantage that they use volatile solvents, they are unable to resolve InsP_4s, InsP_5s and InsP_6 effectively (and with some batches of resin even InsP_3 and InsP_4 prove problematic). We have developed a protocol using open-topped columns of Bio-Rad AG 1.8 (200–400 mesh in the chloride form) anion exchange resin and HCl as an eluant to separate ^{3}H-labelled inositol phosphates contained in enzyme assays or cellular extracts. To separate Ins$P$ + InsP_2 from InsP_3, InsP_4 and InsP_5 + InsP_6 we use 2.9 cm × 0.8 cm of resin contained in Kontes Disposaflex columns (supplied by Hoeffer Scientific Instruments UK Ltd, Newcastle Under Lyme). The aqueous extracts (from e.g. KOH or tri-*n*-octylamine/Freon neutralizations, see Chapter 2) are loaded in 5 ml of H_2O, washed with a further 10 ml of H_2O then eluted with 10 ml (in single batches) of 40 mM HCl (Ins$P$ + InsP_2), 170 mM HCl (InsP_3) 325 mM HCl (InsP_4) and 1.5 M HCl (InsP_5 + InsP_6). All of these HCl solutions can be mixed with an equal volume of any standard, H_2O-compatible scintillation fluid for counting. The columns can be reused (after being washed twice with 10 ml of H_2O) at least 20 times. The principal advantage of the protocol described above is that an average of only 1% of the InsP_3 elutes in the InsP_4 fraction whilst 2–3% of the InsP_4 elutes in the InsP_3 fraction. Also, the HCl eluant is likely to be compatible with the mass determination assay described by Maslanski and Busa (see Chapter 9).

InsP_5s and InsP_6 are much more difficult to resolve but they can be completely separated (an average of less than 0.8% spillover either way) using 12.0 cm × 0.8 cm columns of resin (resin and column supports as described above). D/L-Ins(1,2,3,4,5)P_5 and Ins(1,2,3,4,6)P_5 are eluted by five sequential washes with 10 ml of 0.65 M HCl (each taking about 1 h), D/L-Ins(1,2,4,5,6)P_5 by six 10 ml washes with 0.65 M HCl and Ins(1,3,4,5,6)P_5 by eight 10 ml batches of 0.63 M HCl. InsP_6 is eluted with two 10 ml batches of 1.5 M HCl.

CONCLUSIONS

PEI-cellulose chromatography is a simple and efficient method for the separation of inositol phosphates, enabling multiple samples (at least 20 per plate) to be analyzed in 1.5–2 h. Preliminary results using phytic acid hydrolysate suggest that different isomers of InsP_3 and InsP_4 have slightly different mobilities in 0.5 M HCl and we hope to be able to exploit this further, for example to separate Ins(1,3,4)P_3 from Ins(1,4,5)P_3. If this is possible, PEI-cellulose chromatography will provide not only a rapid alternative to Dowex columns, but also could be a satisfactory replacement for some high-performance liquid chromatography (HPLC) applications.

Anion exchange column chromatography using resin in the chloride form also provides an extremely precise method for separating the higher inositol phosphates, and is less time-consuming than HPLC for analysing multiple samples.

Acknowledgments. C.E.L.S. is supported by an MRC Research Studentship. L.R.S. is a Babraham Fellow. We thank Phill Hawkins for advice and encouragement.

REFERENCES

1. Downes CP, Michell RH. The polyphosphoinositide phosphodiesterase of erythrocyte membranes. *Biochem J* 1981;198:133–140.
2. Emilsson A, Sundler R. Differential activation of phosphatidylinositol deacylation and a pathway via diphosphoinositide in macrophages responding to zymosan and ionophore A23187. *J Biol Chem* 1984;259:3111–31116.
3. Ryu SH, Lee SY, Lee KY, Rhee SG. Catalytic properties of inositol trisphosphate kinase: activation by Ca^{2+} and calmodulin. *Fed Am Soc Exp Biol J* 1987;1:389–393.
4. Clarke NG, Dawson RMC. $O \rightarrow N$-transacylation. A new method for the quantitative deacylation of phospholipids. *Biochem J* 1981;195:301–306.
5. Berridge MJ, Dawson RMC, Downes CP, Heslop JP, Irvine RF. Changes in the levels of inositol phosphates after agonist-dependent hydrolysis of membrane phosphoinositides. *Biochem J* 1982;212:473–482.

Methods in Inositide Research,
edited by Robin F. Irvine.
Raven Press, Ltd., New York © 1990

5

Ion-Pair Chromatography: A Method for the Separation of Inositol Phosphates after Labelling of Cells with [^{32}P]Phosphate

Jean-Claude Sulpice*, Christilla Bachelot**, Philippe Gascard*, and Françoise Giraud*

** Physiologie de la Nutrition, Université Paris XI, Bâtiment 447, CNRS UA 646, 91405 Orsay Cedex, France; and ** INSERM U150, Hôpital Lariboisière, 75475 Paris, France*

The most commonly used method to measure the production of inositol phosphates (Ins*P*s) is to label their direct precursors, the phosphoinositides, from [^{3}H]inositol. In this way only these species of molecules are labelled, facilitating the chromatographic analysis of the cytosolic extracts. However, in some cells, such as platelets, [^{3}H]inositol incorporation requires long incubation periods and/or permeabilization treatments, leading to uncontrolled activation or partial loss of the responses to agonists (1,2). In addition, [^{3}H]inositol cannot be incorporated into phospholipids of mammalian erythrocytes. The cellular incorporation of [^{32}P]P_i, being a more efficient and less expensive way to label the phosphoinositides, leads to a great number of highly labelled phosphorylated metabolites, among which Ins*P*s account for only a small percentage of the total radioactivity.

No simple method, with the exception of high-voltage electrophoresis (3), has been proposed for the separation of [^{32}P]Ins*P*s from the other ^{32}P-labelled metabolites. Ion exchange (high-performance liquid chromatography, HPLC) with strong (SAX) (4,5) or weak (WAX) anion exchange (6,7) does not allow the separation of nucleotides from Ins*P*s, especially ATP from InsP_3. Thus, the analysis of [^{32}P]Ins*P*s requires various chromatographic or enzymatic pre-treatments to remove the nucleotides 2,3-bisphosphoglycerate (2,3-DPG) and pyrophosphate (PP_i) (1,2,8,9). Ion-pair chromatography offers an interesting alternative in the separation of both organic acids, e.g. nucleotides (10), and bases.

In this chapter, we discuss the application of this chromatographic method to the quantitative analysis of labelled Ins*P*s, after [^{32}P]P_i (or-[^{3}H] inositol) labelling.

MATERIALS AND METHODS

Materials

myo-Inositol (1,3,4,5,6)-pentakisphosphate (Ins(1,3,4,5,6)P_5) and *myo*-[^{3}H]inositol hexakisphosphate (InsP_6) were generous gifts from Drs L. R. Stephens and R. F. Irvine. *myo*-[^{3}H]InsP_6 was obtained also from CEA (France). ^{32}P-labelled *myo*-inositol (1,4,5)-trisphosphate (Ins(1,4,5)P_3), *myo*-[2-^{3}H]inositol (1,3,4,5)-tetrakisphosphate (Ins(1,3,4,5)P_4), *myo*-[2-^{3}H]Ins-(1,4,5)P_3, [^{14}C]inositol phosphate (Ins*P*) and *myo*-[2-^{3}H]inositol were obtained from Amersham (France) and [^{32}P]P_i from International CIS (France).

ATP, ADP, AMP, GTP, 2,3-DPG, fructose 6-phosphate and 1,6-bisphosphate, glucose 6-phosphate and 1,6-bisphosphate, Ins(1,4,5)P_3 and phytic acid were obtained from Sigma (St Louis, MO).

Cell Labelling and Stimulation

The production of Ins*P*s has been studied in different types of cells after labelling with [^{32}P]P_i and stimulation with agonists or Ca^{2+}. The level of cell labelling is a critical parameter for improving the threshold of detection of the labelled Ins*P*s. Conditions for [^{3}H]inositol uptake in hepatocytes and for its incorporation into phosphoinositides have been discussed by Prpic et al. (11). [^{32}P]P_i incorporation into ATP and polyphosphoinositides is dependent on the types of cells and/or on the saline composition of the incubation medium (12). The labelling conditions that we have used are described in Table 1.

Some difficulties in the separation procedure can arise from the composition of the final medium used for cell stimulation. In the presence of high concentrations of divalent cations, the peaks of Ins*P*s become doublets. When the Ca^{2+} or Mg^{2+} concentration in the incubation medium is higher than 2 mM, we advocate centrifugation of the cell suspensions in order to remove the medium before quenching of the reaction with $HClO_4$ (see below). Several classical physiological media which can be used without any problem are described in Table 2.

Sample Treatment for HPLC Analysis

One volume of cell suspension was mixed with 0.5 volumes of 10% $HClO_4$ (15 min at 4°C). [^{3}H]Ins(1,4,5)P_3 and [^{3}H]Ins(1,3,4,5)P_4 (2000 c.p.m. each) were added for internal standardization and phytic acid hydrolysate (25 μg of phosphorus) (13) as carrier for Ins*P*s. The samples were centrifuged (10 min, 1500 g.). The $HClO_4$ extract was neutralized with saturated $KHCO_3$ (pH 7.5–8) and centrifuged. The supernatant was adjusted to 800 μl with

TABLE 1. *Conditions for [^{32}P]P$_i$ labelling*

Cells	Cell concentration (cells/ml)	Medium composition	[^{32}P]P$_i$ radioactivity (MBq/ml)	Time and temperature of incubation
Human erythrocytes	4×10^9	150 mM NaCl, 10 mM glucose, 1.5 mM Hepes (pH 7.4)	0.74	150 min, 37°C
Rat liver hepatocytes	6×10^{10}	116 mM NaCl, 5.4 mM KCl, 0.92 mM NaH_2PO_4, 25 mM $NaHCO_3$, 0.81 mM $MgCl_2$, 1.8 mM $CaCl_2$, 5 mM glucose, amino acids, vitamins, under O_2/CO_2 (19:1) (Eagle's medium)	0.41	90 min, 37°C
Human platelets	PRP (8×10^5)	Plasma	37	90 min, 37°C

Erythrocytes were prepared as described by Rhoda et al. (31) and hepatocytes as described by Mauger et al. (37). PRP (platelet-rich plasma) was obtained as described by Levy-Toledano et al. (38). After incubation with [^{32}P]P$_i$ cells were washed three times in the incubation medium (or in 140 mM NaCl, 5 mM KCl, 12 mM trisodium citrate (pH 6), 10 mM glucose, 12.5 mM sucrose for platelets) and resuspended in the medium used for cell stimulation (see Table 2).

H_2O and mixed with 100 mg of charcoal (Darco, type G-60). Samples were left for 5 min on ice and centrifuged (1 min, 12,000 *g*). To about 500 μl of the supernatant, [^{3}H]Ins(1,3,4)P_3 prepared as described below and 100 μM AMP (100 μl) were added and mixed with 350 μl of the HPLC eluant. The pH was adjusted to 6.5–7 and the sample was injected though a Rheodyne injector with a 1 ml loop, with a syringe adjusted on a Dinagard filter, 0.2 μm PP (Microgon Inc., Laguna Hills, CA).

Preparation of Labelled Ins*P*s Standards

[^{3}H]Ins(1,3,4)P_3 and [^{3}H]Ins(1,3,4,5)P_4 were prepared from [^{3}H]Ins(1,4,5)P_3 using a soluble fraction of rat liver homogenate according to the methods of Irvine et al. (14) and Shears et al. (15), with the following modification. The reaction was processed for 45 min for Ins(1,3,4,5)P_4 and for 75 min for Ins(1,3,4)P_3 and quenched with 10% (v/v) ice-cold $HClO_4$. [^{3}H]Ins(1,3,4)P_3 (and InsP_2) can be prepared also with a better yield from authentic [^{3}H]Ins(1,3,4,5)P_4, as described by Doughney et al. (16). [^{32}P]Ins(1,4,5)P_3 was prepared from ^{32}P-prelabelled erythrocyte membranes (17). ^{32}P-labelled glycerophosphate (Gro*P*), glycerophosphoinositol 4-phosphate (Gro*P*Ins(4)*P*) and 4,5-bisphosphate (Gro*P*Ins(4,5)P_2) were prepared by mild

TABLE 2. *Conditions for cell stimulation*

Cells	Cell concentration (cells/ml)	Cell volume per assay (ml)	Medium	$[Mg^{2+}]$	$[Ca^{2+}]$	Stimulus	Time of stimulation
Erythrocytes	10^9	0.3	75 mM NaCl, 75 mM KCl, 10 mM Hepes (pH 7.4)	0.15 mM	10 μM to 1 mM	A23187, 5 μM	10 min
Hepatocytes	17×10^6	0.5	Eagle's medium	0.81 mM	1.8 mM	Vasopressin, 0.2 μM	5–60 s
Platelets	5×10^8	0.4	140 mM NaCl, 3 mM KCl, 5 mM $NaHCO_3$, 10 mM glucose, 10 mM Hepes (pH 7.4)	5 mM	0	Thrombin, 2 units/ml	3–120 s

A23187 and vasopressin were obtained from Sigma and thrombin from Roche (Basel, Switzerland).

alkaline hydrolysis of ^{32}P-labelled phosphatidic acid and phosphoinositides from erythrocyte membranes as described (18).

Identification of the ^{32}P-labelled Metabolites

Nucleotides (*ATP, ADP, AMP* and *GTP*)

These were detected by ultraviolet absorption and identified by comparison of their retention time with those of authentic standards.

2,3-DPG

The neutralized extract from ^{32}P-labelled erythrocytes was treated with phosphoglycerate mutase (rabbit muscle, 5400 units/ml, Sigma) as described (12,19) and analysed directly by HPLC. The profile obtained was compared with that of an untreated sample.

Fructose 1,6-bisphosphate

The neutralized extract from ^{32}P-labelled erythrocytes (300 μl) was adjusted to pH 9.5 with 2 M KOH and to 1 mM $MgCl_2$. Fructose 1,6-bisphosphatase (rabbit muscle, Sigma) suspension (34 μl, 2 units) was added and the mixture incubated (30 min, 25°C). After centrifugation (1 min, 12,000 *g*), the supernatant was analysed by HPLC. The profile obtained was compared with that of an untreated sample.

Inositol Phosphates

[^{32}P]Ins*P*s were identified from their co-elution with the standards, either authentic [^{3}H]Ins(1,4,5)P_3 and [^{3}H]Ins(1,3,4,5)P_4 or [^{3}H]Ins(1,3,4)P_3 prepared as described above. [^{3}H]Ins*P*s were identified from their elution times and by co-elution with authentic [^{32}P]Ins(1,4,5)P_3 for the [^{3}H]Ins(1,4,5)P_3. Further identification was carried out as follows: HPLC fractions containing the [^{3}H]Ins(1,4,5)P_3 and the corresponding ^{32}P radioactivity were pooled (3 ml), treated with 1 ml 10% ice-cold $HClO_4$ and centrifuged (10 min, 3000 *g*). Methanol (4 ml) was added to the supernatant and the mixture was centrifuged again. The supernatant was dried, the residue redissolved in water (1.2 ml), and the sample centrifuged (1 min, 12,000 *g*). The last supernatant was injected into a Partisil 10-SAX HPLC column, eluted with H_2O for 3 min and, subsequently, by three successive convex gradients (Waters Millipore Corp. gradient programs 4, 2 and 1) of increasing ammonium formate concentration (0–1.7 M) adjusted to pH 3.7 with H_3PO_4 (4,5).

[^{32}P]Ins(1,4,5)P_3 was found to co-migrate with [^{3}H]Ins(1,4,5)P_3 in this reference chromatographic system and the ^{32}P:^{3}H radioactivity ratio was not changed.

HPLC Analysis

The HPLC system (Waters, Millipore Corp., Milford, MA) consists of two pumps (model 501), a gradient controller (model 480) and a 481 spectrophotometer or R401 Refraction Index detector. The radioactive flow detector (Flo-One HP/S Radiomatic Instruments, FL) is used only in optimization assays. Several octadecyl reverse-phase columns have been tested for ion-pair chromatography. Lichrosorb RP18 (5 μm) 25 cm × 0.4 cm (Merck, West Germany), μBondapack C18 30 cm × 0.4 cm (Waters, Millipore Corp., MA), Sup. RS classic 25 cm × 0.46 cm (Prolabo, Paris, France) and Ultrasphere IP (5 μm) 25 cm × 0.46 cm (Beckman, USA). All these columns had the same ability to separate Ins*P*s from nucleotides. However, the non-end-capped phases (Lichrosorb RP18 and μBondapack C18) rapidly led to a loss of retention capacity. The Ultrasphere IP has been used for several months (about 500 runs). Partisil 10 SAX 25 cm × 0.46 cm (Whatman, USA) was used for ion exchange chromatography. Reagents used for the preparation of elution solvents are: H_2O purified on Milli Q apparatus (Waters, Millipore Corp., MA), acetonitrile HPLC grade, KH_2PO_4 (Prolabo, Paris, France), tetrabutylammonium hydrogensulphate (TBAHS) for synthesis grade (Merck, West Germany). Before use, the solvents were filtered and the pH adjusted with 6 M KOH.

The eluant flow was always 1 ml/min and the ultraviolet-detection of nucleotides was performed at 254 nm. Collected fractions (0.5 or 1 ml) were mixed with 5 ml of scintillation liquid (Aqualuma-plus, Lumac, the Netherlands). ^{32}P radioactivity was determined in c.p.m. at a constant efficiency, with a two-channel liquid scintillation spectrometer MR300 (Kontron, France). The window setting ^{3}H/^{32}P induced the transfer of only 1% of the ^{32}P counts into the ^{3}H window.

Calculations

The total radioactivity of each inositol phosphate is obtained from the sum of the counts (minus baseline background) in the fractions that co-eluted with the corresponding ^{3}H-labelled standards (InsP_3, InsP_4) and corrected for recovery (about 50%, calculated from [^{3}H]Ins*P* internal standards added at the extraction step).

RESULTS AND DISCUSSION

Optimization of the Chromatographic System

Ion-pair chromatography is more versatile than ion exchange chromatography, since it is affected by more parameters. The choice of the counterion and of its concentration, of the salts, of the organic solvents and of the pH of the elution buffer offers many possibilities to optimize the separation conditions. In this work, the counterion was TBAHS. Indeed tetrabutyl ammonium (TBA) is the most generally used counterion, the sulphate form being the salt with the smallest degree of ultraviolet absorption. Shayman and BeMent (20) have tested several quaternary ammonium compounds with different carbon chain lengths. They have also chosen TBA (at the concentration of 50 mM). In this study, we have used TBAHS (25 mM).

Figure 1 shows the respective mobility of nucleotides and Ins(1,4,5)P_3 under different conditions. The ionic strength and the pH were first set at constant values (100 mM KH_2PO_4, pH 5.5) to investigate the effect of variation of acetonitrile concentration (Fig. 1a). At 0% acetonitrile, Ins(1,4,5)P_3 eluted after AMP and before ADP and ATP, whereas at 20% (v/v) acetonitrile it eluted after the nucleotides. In this second situation (20% acetonitrile), decreasing the ionic strength to 20 mM KH_2PO_4 (Fig. 1b) or increasing the pH to 7 (Fig. 2a), improved the resolution of Ins(1,4,5)P_3 from ATP, 20% acetonitrile being the optimum value, with a retention time for Ins(1,4,5)P_3 of less than 30 min. However, at this pH, divalent cations and organic buffers coming from the cell incubation medium led to considerable disturbance in the retention times and shape of the peaks (which became doublets).

Therefore this isocratic system was not used further and a two-step elution scheme was adopted: first, elution of Ins(1,4,5)P_3 after AMP with solvent A [25 mM TBAHS, 100 mM KH_2PO_4(pH 5.5)] for 40 min and, secondly, elution of ADP, ATP with solvent A/acetonitrile (80:20, v/v) over the next 15 min. These conditions have been used for the direct analysis of the cytosolic extracts of erythrocytes without elimination of ATP, which eluted after InsP_3 (Fig. 3). Under the same chromatographic conditions, the separation of [^{3}H]Ins(1,3,4)P_3 standard from [^{32}P]- or [^{3}H]Ins(1,4,5)P_3 standard was possible, but the [^{3}H]Ins(1,3,4,5)P_4 standard co-eluted with ATP. The separation of Ins(1,3,4,5)P_4 from ATP was then achieved as described for separating Ins(1,4,5)P_3 from ATP (see Fig. 1b). At pH 5.5 and 20% acetonitrile, decreasing the KH_2PO_4 concentration to 10 mM led to a good resolution of ATP and InsP_4. The optimal conditions were: solvent A /acetonitrile (99:1, v/v) for 28 min, followed by 25 mM TBAHS, 10 mM KH_2PO_4 (pH 5.5)/acetonitrile (77:23, v/v) over the next 22 min. A 5 min re-equilibration step was necessary before a new run was started.

In the isocratic system used by Shayman and BeMent (20) (0.05 M tetrabutyl ammonium phosphate (TBAP), 0.04 M KH_2PO_4/acetonitrile (81:19,

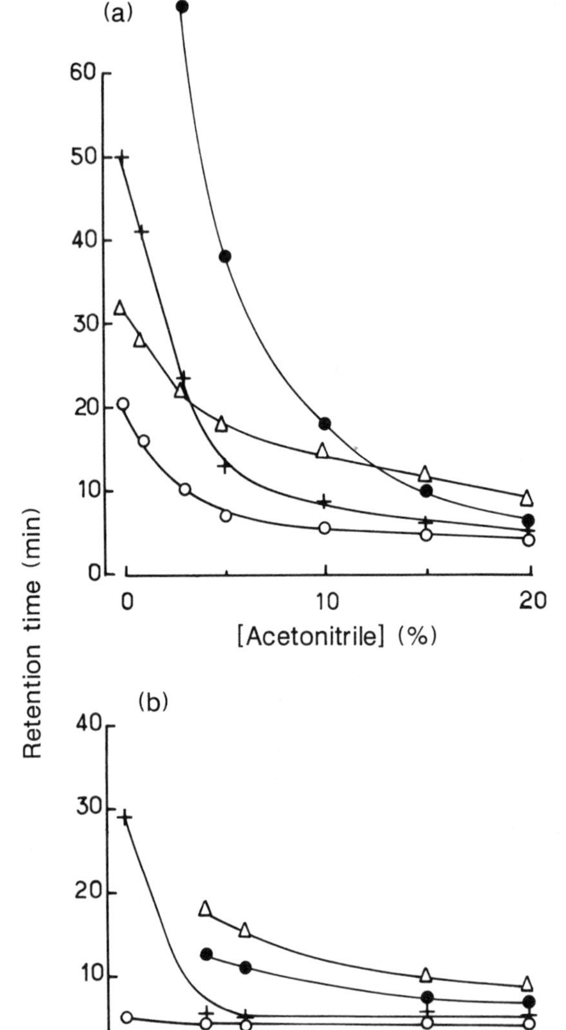

FIG. 1. Separation of Ins(1,4,5)P_3 and adenosine nucleotide standards by ion-pair HPLC. **(a)** Dependence on the percentage of acetonitrile at high ionic strength (100 mM KH_2PO_4, 25 mM TBAHS (pH 5.5)). **(b)** Dependence on the concentration of KH_2PO_4 (20% acetonitrile, 25 mM TBAHS (pH 5.5)). △, Ins(1,4,5)P_3; ●, ATP; +, ADP; ○, AMP. In the absence of KH_2PO_4, ATP and Ins(1,4,5)P_3 had not eluted by 40 min. (Reproduced with permission from ref. 32.)

v/v), pH 3.5), elution times of ATP and InsP_3 were about 8 and 20 min, respectively. Under similar conditions at pH 5.5 (Fig. 1b), elution times of ATP and InsP_3 were close to those observed by Shayman and BeMent; however, Ins(1,3,4)P_3 eluted before Ins(1,4,5)P_3, whereas the converse was observed unexpectedly in the former experiment (20).

In principle, an isocratic system such as that described above or that used

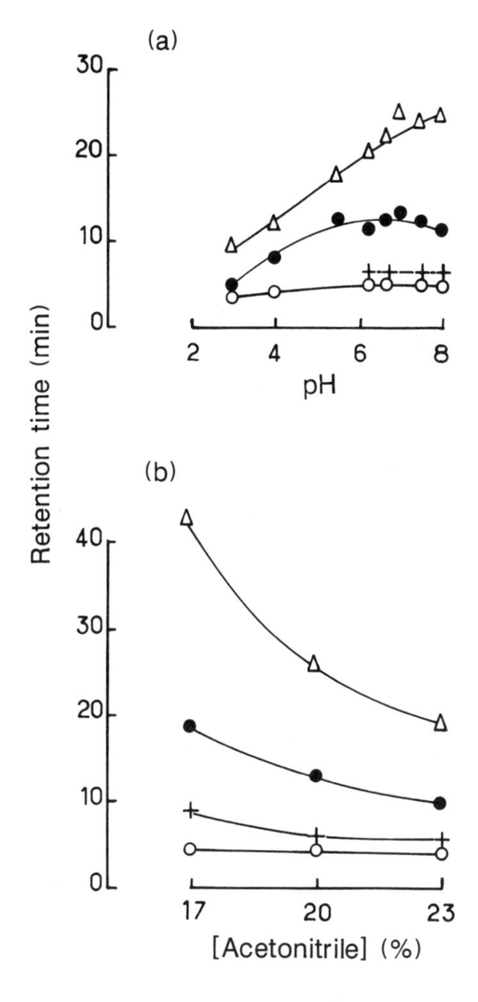

FIG. 2. Separation of Ins(1,4,5)P_3 and adenosine nucleotide standards by ion-pair HPLC. (a) Dependence on the pH (25 mM TBAHS, 20 mM KH_2PO_4, 20% acetonitrile). (b) Dependence on the percentage of acetonitrile (25 mM TBAHS, 20 mM KH_2PO_4 (pH 5.5)). △, Ins(1,4,5)P_3; ●, ATP; +, ADP; ○, AMP. (Reproduced with permission from ref. 32.)

by Shayman and BeMent (20) offers obvious advantages. However, (i) after ^{32}P-labelling, prior elimination of ATP is required to avoid interference between the ^{32}P background following the ATP peak and the small peaks of Ins*P*s, and (ii) in all isocratic systems tested, the elution of InsP_5 and InsP_6 in less than 80 min did not allow a clear separation of InsP_2 from InsP_3. An acetonitrile gradient led to a good separation of these two latter InsPs- but not to that of InsP_5 and InsP_6. To optimize these separations, a KH_2PO_4 gradient was developed. Solvent I was 25 mM TBAHS, 40 mM KH_2PO_4/acetonitrile (85:15, v/v) (pH 5.5) and solvent II was 25 mM TBAHS, 150 mM KH_2PO_4/acetonitrile (85:15, v/v) (pH 5.5). Elution was (i) 100% I (15 min), (ii) a linear gradient from 100% I to 100% II (25 min), (iii) 100% II (30 min). Under these conditions the elution times were 21, 39, 52 and 60

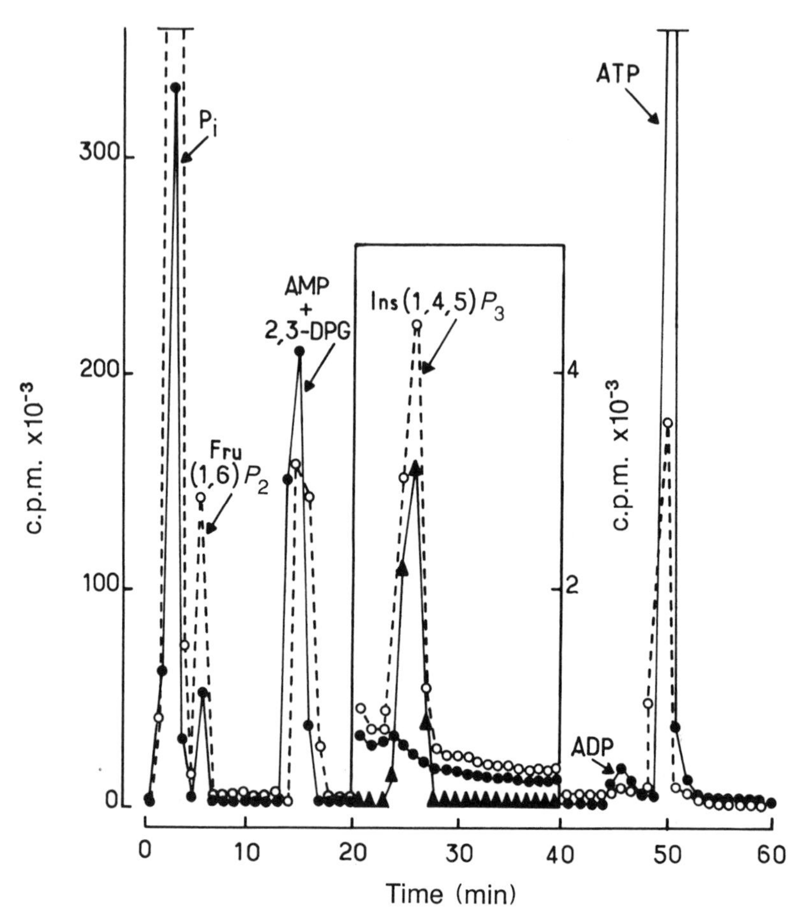

FIG. 3. HPLC profile of ^{32}P-labelled compounds from cytosolic extracts of erythrocytes. Erythrocytes, prelabelled with [^{32}P]P_i, were incubated for 20 min in a saline medium containing 5 μM A23187 and either 1 mM EGTA (●) or 1 mM $CaCl_2$ (○). A sample of the neutralized extract was applied to the column and elution was with solvent A (see the text) for 42 min and, with solvent A/acetonitrile (80:20, v/v) over the next 18 min (see the text). ●, ○, ^{32}P; ▲, [^{3}H]Ins(1,4,5)P_3 standard. Further identification of Ins(1,4,5)P_3 in the Ca^{2+}-stimulated sample was carried out as follows: fractions eluted between 23 and 29 min were collected, desalted, freeze-dried and reinjected into a Partisil 10-SAX HPLC column. [^{3}H]Ins(1,4,5)P_3 standard was found to co-elute with a ^{32}P peak in this second system. (Reproduced with permission from ref. 32.)

min, respectively, for Ins(1,4,5)P_3, Ins(1,3,4,5)P_4, Ins(1,3,4,5,6)P_5 and InsP_6, the peak corresponding to the last of these being very broad (results not shown).

The mechanism of ion-pair chromatography can be explained by ion-pair formation in the mobile phase, ion exchange at the stationary phase surface or dynamic equilibrium (21). Arguments in favour of the ion-exchange mechanism are that: (i) high concentrations of the counterion (25 mM TBAHS) and a high percentage of cross-linking of the reverse-phase lead to counterion adsorption; (ii) retention times of the phosphorylated metabolites depend on the number of the phosphate groups. This mechanism has been postulated by Shayman and BeMent (20). Some of our results show that the mobility of the nucleotides was affected more by the acetonitrile concentration, whereas the mobility of the inositol phosphates was affected more by the ionic strength. This could indicate that ion-pair chromatography cannot be explained by only a simple anion-exchange model.

Sensitivity of the Measurement of Radioactive Ins*P*s

Different parameters can affect the threshold of detection of each radioactive peak. The first is the ^{32}P or ^{3}H background level, which is generally less than 50 c.p.m. in each fraction. Ins*P*s in the neutralized extract at −20°C are very stable, but the stability of ATP, before charcoal treatment, is a critical point. Degradation of a small percentage of ATP will lead to a more elevated background, preventing proper detection of the small Ins*P*s peaks. The second important parameter is the possible adsorption of the phosphate groups on the polar supports (glassware, charcoal, tubing and column) that will lead to losses of Ins*P*s. In particular, the anion exchange properties of the charcoal (22) result in Ins*P*s losses: 95% at pH < 7 and only 17% and 4–5%, respectively, for InsP_4 and InsP_3 at pH > 7.5. The addition of phytic acid hydrolysate (25 μg of phosphorus) (13) totally suppressed the loss of Ins*P*s at this step, whereas the addition of mannitol (12.5 mM) (4) had no effect. When phytic acid hydrolysate was added at the extraction step, the final recovery was about 50%. The major causes of loss were the dead volume of the charcoal and that of the filter.

Production of Ins(1,4,5)P_3 in Ionophore-Treated Erythrocytes

The erythrocyte membrane phospholipase C can be activated by Ca^{2+} (17,23). Treatment of intact erythrocytes with the ionophore A23187 and Ca^{2+} induces a breakdown of polyphosphoinositides (24–26). However, the production of Ins(1,4,5)P_3 has never been investigated under these conditions. ^{32}P-labelled erythrocytes were treated with 5 μM A23187, 1 mM $CaCl_2$ or 1 mM EGTA, and the ^{32}P-labelled metabolites in the cytosolic extracts were analysed (Fig. 3). In the presence of Ca^{2+}, the ATP peak was drastically reduced as a consequence of the activation of the ATP-dependent Ca^{2+} pump. At 30 μM external Ca^{2+}, the maximal production of Ins(1,4,5)P_3 was reached between 10 and 20 min, whereas at lower doses of Ca^{2+}, Ins(1,4,5)P_3 was formed more slowly (Fig. 4a). When external Ca^{2+} was increased from 10 μM to 1 mM, Ins(1,4,5)P_3 production increased

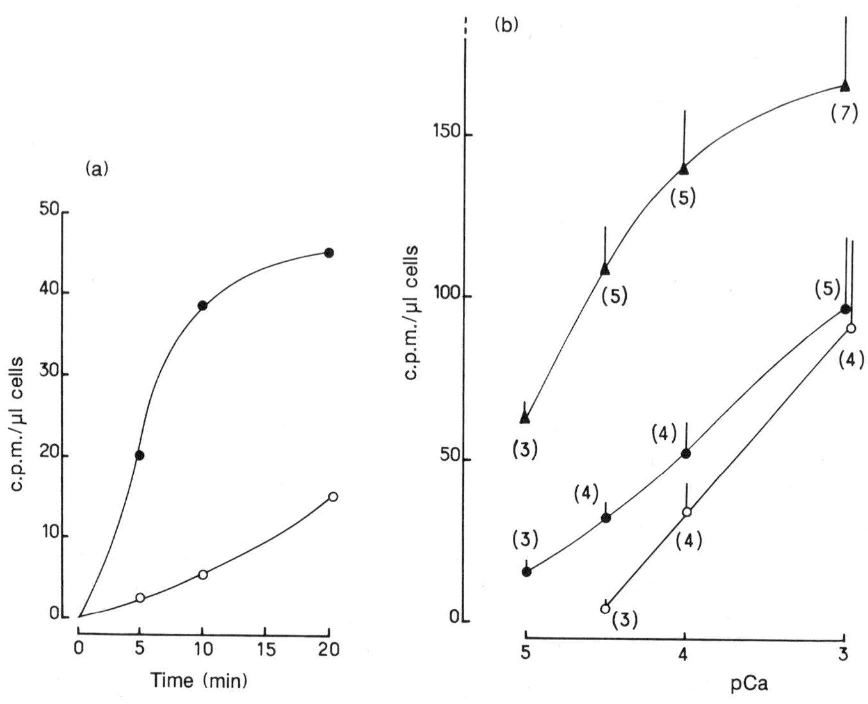

FIG. 4. (a) Time course of Ins(1,4,5)P_3 production in erythrocytes. ^{32}P-prelabelled erythrocytes were incubated in a saline medium containing 5 µM A23187 and either 15 µM (○) or 30 µM (●) $CaCl_2$. Radioactivity in Ins(1,4,5)P_3 was measured by HPLC as indicated in the legend to Fig. 3. (Reproduced with permission from ref. 32.) **(b)** Ca^{2+} and Mg^{2+} dependence of Ins(1,4,5)P_3 production and of PtdIns(4,5)P_2 breakdown in erythrocytes. ^{32}P-prelabelled erythrocytes were incubated in a saline medium containing 5 µM A23187 and 10 µM to 1 mM $CaCl_2$ with either 0.15 mM $MgCl_2$ (●, ▲) or 1.2 mM $MgCl_2$ (○). Radioactivity in Ins(1,4,5)P_3 (●, ○) was measured by HPLC as indicated in the legend to Fig. 3. Radioactivity in PtdIns(4,5)P_2 was measured after acidic extraction of the washed membrane and thin-layer chromatography of the lipid extract (31). [^{32}P]PtdIns(4,5)P_2 breakdown (▲) was calculated as the difference between the radioactivity of PtdIns(4,5)P_2 in the presence of A23187 and 1 mM EGTA and that in the presence of A23187 and $CaCl_2$. Values are means ± S.E.M. (number of experiments is indicated in parentheses). (Reproduced with permission from ref. 32.)

linearly with no indication of saturation (Fig. 4b). The increase in intracellular free Mg^{2+} concentration from 0.4 mM (the normal physiological value obtained by adding 0.15 mM to the medium) to 2 mM (obtained by adding 1.2 mM to the medium) (27) induced a decrease in Ins(1,4,5)P_3 production, particularly at low Ca^{2+} concentration. This effect did not result from an inhibition of phospholipase C by Mg^{2+} because, in parallel experiments, phosphatidylinositol 4,5-bisphosphate (PtdIns(4,5)P_2) breakdown was shown to be independent of this cation (results not shown). This decrease in Ins(1,4,5)P_3 production is attributed to the Mg^{2+} activation of

Ins(1,4,5)P_3 phosphomonoesterase. In Fig. 4b, the decrease in the radioactivity of PtdIns(4,5)P_2 as a function of Ca^{2+} is also shown. Clearly, this decrease was greater than the increase in the radioactivity of Ins(1,4,5)P_3 at each Ca^{2+} concentration tested. This difference can be explained by the simultaneous activity of phospholipase C and of Ins(1,4,5)P_3 phosphomonoesterase. These data obtained in intact erythrocytes validate previous observations taken from experiments carried out in isolated membranes (17,23,28). In addition it shows that subtle changes in [^{32}P]Ins(1,4,5)P_3 content can be detected in cells despite the high levels of ^{32}P-labelled nucleotides.

Production of Ins*P*s in Vasopressin-Stimulated Hepatocytes and Thrombin-Stimulated Platelets

The extracts of hepatocytes and platelets, in comparison to those of erythrocytes contain a greater number of ^{32}P-labelled metabolites, mainly nucleotides. The chromatographic method used in this study does not require, in principle, elimination of the major nucleotides AMP, ADP and ATP, because they do not interfere with inositol phosphates. However, due to their high radioactivity relative to that of the inositol phosphates, their removal by charcoal treatment – which also eliminated other minor contaminants – resulted in an improvement of the chromatographic data. As shown in Figs. 5 and 6, the removal of ATP and ADP was not complete, contrasting with the observation of Meek (8), obtained using ultraviolet recording.

Figure 5 shows the elution profile of ^{32}P-labelled metabolites in a cytosolic extract from hepatocytes either unstimulated (Fig. 5a) or after stimulation with vasopressin (Fig. 5b). The very low ^{32}P background level in the region of [^{3}H]Ins*P*s standards in control cells and the appearance of ^{32}P peaks comigrating with the ^{3}H standards in stimulated cells demonstrated that these peaks could be attributed to the Ins*P*s produced on receptor activation. Similar conclusions could be reached with unstimulated platelets (Fig. 6a) and thrombin-activated platelets (Fig. 6b).

Figure 7a shows the time courses of the production of inositol phosphates in ^{32}P-labelled hepatocytes after stimulation with vasopressin. Ins(1,4,5)P_3 production was maximal between 5 and 30 s, whereas Ins(1,3,4,5)P_4 and Ins(1,3,4)P_3 were formed more slowly. These data are in agreement with previous results obtained in [^{3}H]inositol-labelled hepatocytes (5,29,30). Figure 7b shows the time course of the production of Ins*P*s in ^{32}P-labelled platelets after stimulation with thrombin. Ins(1,4,5)P_3 production was maximal after 5 s and its conversion into Ins(1,3,4,5)P_4 and Ins(1,3,4)P_3 was much more rapid than in stimulated hepatocytes. Ins(1,3,4,5)P_4 also disappeared quickly, whereas Ins(1,3,4)P_3 production was very large and its

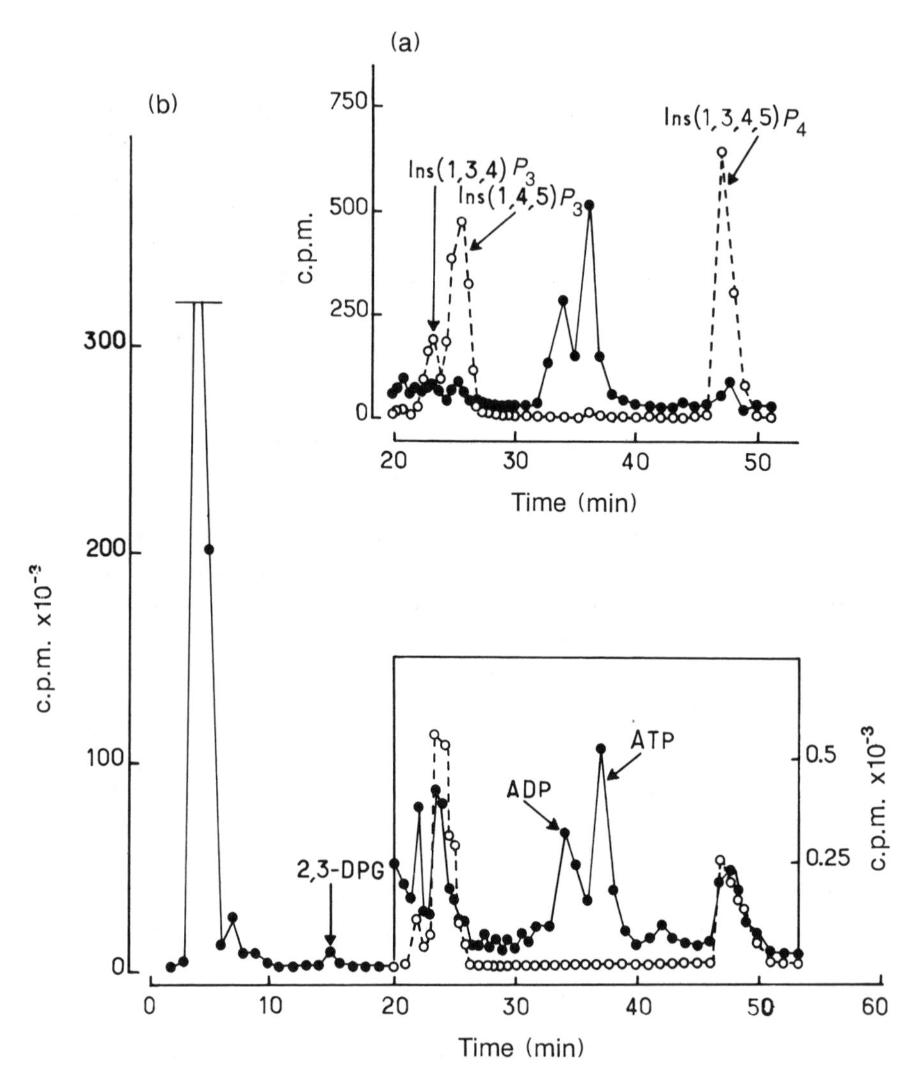

FIG. 5. HPLC profiles of ^{32}P-labelled compounds from a cytosolic extract of **(a)** unstimulated and **(b)** vasopressin-stimulated hepatocytes. Hepatocytes, prelabelled with $[^{32}P]P_i$, were incubated for 30 s with 0.2 μM vasopressin (see Tables 1 and 2 for experimental details). The neutralized extracts were treated with charcoal (see Materials and Methods). Elution was with solvent A/acetonitrile (99:1, v/v) for 28 min and with 25 mM TBAHS, 10 mM KH_2PO_4 (pH 5.5)/acetonitrile (77:23,v/v) over the next 22 min. ●, ^{32}P; ○, ^{3}H-labelled standards. (Reproduced with permission from ref. 32.)

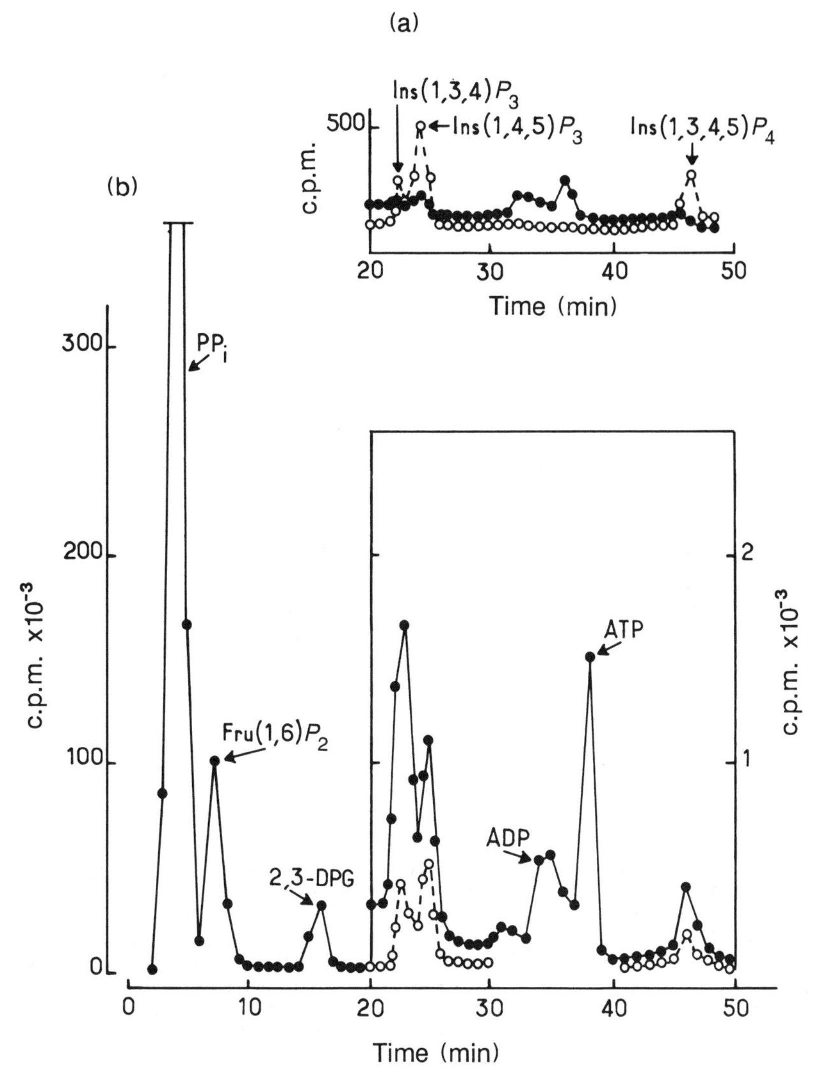

FIG. 6. HPLC profiles of ^{32}P-labelled compounds from cytosolic extracts of (a) unstimulated and (b) thrombin-stimulated platelets. Platelets, prelabelled with [^{32}P]P_i, were incubated for 10 s with 2 units thrombin/ml (see Tables 1 and 2 for experimental details). The neutralized extracts were treated with charcoal. Elution was as indicated in the legend to Fig. 5. ●, ^{32}P; ○, ^{3}H-labelled standards. (Reproduced with permission from ref. 32.)

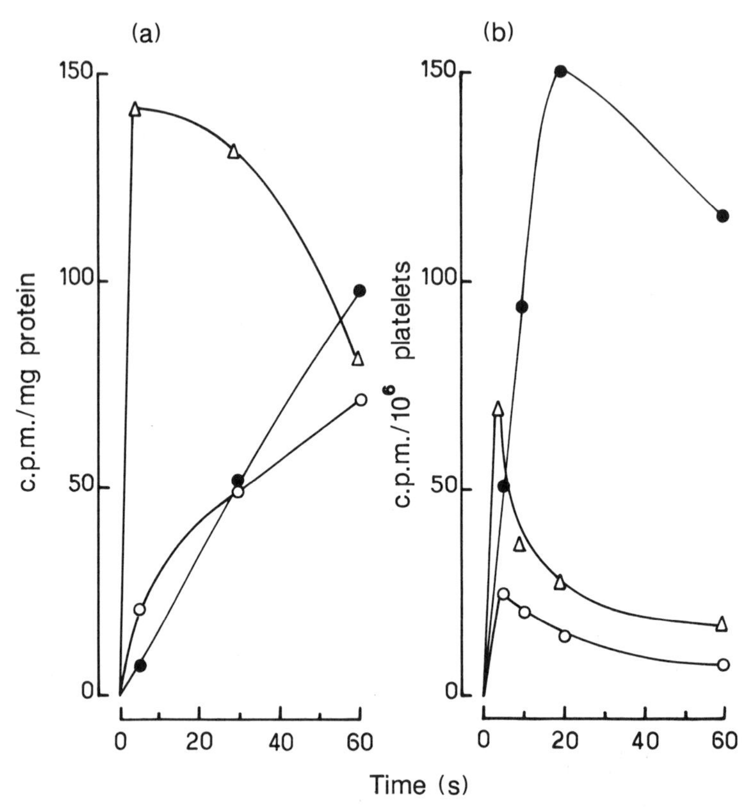

FIG. 7. Time course of Ins*P*s production in (a) hepatocytes and (b) platelets. ^{32}P-prelabelled hepatocytes or platelets were incubated with 0.2 μM vasopression or 2 units thrombin/ml, respectively (see Tables 1 and 2 for experimental details). Ins(1,4,5)P_3 (△), Ins(1,3,4)P_3 (●) and Ins(1,3,4,5)P_4 (○) were measured by HPLC as indicated in the legend to Fig. 5. Values are means of triplicate determinations (hepatocytes) or of duplicate determinations (platelets) in a single experiment. Another experiment gave essentially similar results in hepatocytes and slightly different kinetics (more rapid disappearance of Ins(1,4,5)P_3 and InsP_4) in platelets. To estimate the reproducibiity of the measurements, four samples from one platelet extract were injected sequentially . This gave a statistical variation of the mean of less than 5%. (Reproduced with permission from ref. 32.)

disappearance was slower than that of the other Ins*P*s. These results are in agreement with those of Daniel et al. (9).

The presence of glycerophosphoinositol phosphates in platelets and their separation from Ins*P*s has been reported (6). Under our experimental conditions, Gro*P*Ins(4)*P* was eluted at 7 min and Gro*P*Ins(4,5)P_2 at 25 min. When the Gro*P*Ins*P*s mixture was injected with a sample containing [^{3}H]Ins*P* standards, Gro*P*Ins(4,5)P_2 was eluted in the fractions before Ins(1,3,4)P_3 and was well resolved from the latter (result not shown). The presence of PP_i has been reported in platelet extracts (9). In our chromatographic system, PP_i (Sigma, St Louis, MO) detected with a refractive index detector was eluted at about 4 min and thus cannot interfere with InsP_3 or InsP_4.

CONCLUSIONS

This ion-pair HPLC method, allowing the separate measurement of Ins(1,3,4)P_3, Ins(1,4,5)P_3 and Ins(1,3,4,5)P_4 in ^{32}P-labelled cells (31–33), offers several advantages over previously published methodologies. ^{32}P-labelling, more efficient and less expensive than ^{3}H-labelling, can be used with all types of cells without drastic permeabilization treatment or long incubation periods. The advantage of ion-pair chromatography over ion-exchange is that it provides good resolution of Ins*P*s from PP_i, 2,3-DPG, ADP and ATP. After a simple charcoal pre-treatment to remove ^{32}P-labelled contaminants co-migrating with Ins*P*s, their quantification by this method was possible, even with incomplete elimination of ADP and ATP.

Ion-pair HPLC compared to traditional ion exchange (SAX type) did not require high salt concentrations or drastic pH conditions, permitting a good preservation of the column and of the samples. Good resolution, in shorter retention times, was obtained without employing complex gradients and long re-equilibration periods. This ion-pair chromatography can be employed also after [^{3}H]inositol labelling (in this case, the charcoal treatment is not necessary). This method allows study of Ins*P* metabolism using an isocratic HPLC system (20,34). In addition, on the same system with only slight modifications, the nucleotides are well separated and ATP concentration and specific radioactivity can be easily determined (31,35,36).

REFERENCES

1. Ishii H, Connolly TM, Bross TE, Majerus PW. Inositol cyclic trisphosphate (inositol 1,2-(cyclic)-4,5-triphosphate) is formed upon thrombin stimulation of human plateles. *Proc Natl Acad Sci USA* 1986;83:6397–6401.
2. Tarver AP, King WG, Rittenhouse SE. Inositol 1,4,5-trisphosphate and inositol 1,2-cyclic 4,5-trisphosphate are minor components of total mass of inositol trisphosphate in thrombin-stimulated platelets. *J Biol Chem* 1987;262:17,268–17,271.
3. Agranoff BW, Murthy P, Seguin EB. Thrombin-induced phosphodiesterasic cleavage of phosphatidylinositol bisphosphate in human platelets. *J Biol Chem* 1983;258:2076–2078.
4. Irvine RF, Anggard EE, Letcher AJ, Downes CP. Metabolism of inositol 1,4,5-trisphosphate and inositol 1,3,4-trisphosphate in rat parotid glands. *Biochem J* 1985;229:505–511.
5. Hansen CA, Mah S, Williamson JR. Formation and metabolism of inositol 1,3,4,5-tetrakisphosphate in liver. *J Biol Chem* 1986;261:8100–8103.
6. Binder H, Weber PC, Siess W. Separation of inositol phosphates and glycerophosphoinositol phosphates by high-performance liquid chromatography. *Anal Biochem* 1985;148: 220–227.
7. Wilson DB, Bross TE, Sherman WR, Berger RA, Majerus PW. Inositol cyclic phosphates are produced by cleavage of phosphatidylinositols (polyphosphoinositides) with purified sheep seminal vesicle phospholipase C enzymes. *Proc Natl Acad Sci USA* 1985;82:4013–4017.

8. Meek JL. Inositol bis-, tris-, and tetrakis (phosphate)s: analysis in tissues by HPLC. *Proc Natl Acad Sci USA* 1986;83:4162–4166.
9. Daniel JL, Dangelmaier CA, Smith JB. Formation and metabolism of inositol 1,4,5-trisphosphate in human platelets. *Biochem J* 1987;246:109–114.
10. Sellevold OFM, Jynge P, Aarstad K. High performance liquid chromatography: a rapid isocratic method for determination of creatine compounds and adenine nucleotides in myocardial tissue. *J Mol Cell Cardiol* 1986;18:517–527.
11. Prpic V, Blackmore P, Exton J. Myo-inositol uptake and metabolism in isolated rat liver cells. *J Biol Chem* 1982;257:11,315–11,322.
12. King CE, Stephens LR, Hawkins PT, Guy GR, Michell RH. Multiple metabolic pools of phosphoinositides and phosphatidate in human erythrocytes incubated in a medium that permits rapid transmembrane exchange of phosphate. *Biochem J* 1987;244:209–217.
13. Wreggett KA, Howe LR, Moore JP, Irvine RF. Extraction and recovery of inositol phosphates from tissues. *Biochem J* 1987;245:933–934.
14. Irvine RF, Letcher AJ, Heslop JP, Berridge MJ. The inositol tris/tetrakisphosphate pathway demonstration of Ins(1,4,5)P_3 3-kinase activity in animal tissues. *Nature* 1986;320:631–634.
15. Shears SB, Storey DJ, Morris AJ, Cubit AB, Parry JB, Michell RH, Kirk CJ. Dephosphorylation of *myo*-inositol 1,4,5-triphosphate and *myo*-inositol 1,3,4-triphosphate. *Biochem J* 1987;242:393–402.
16. Doughney C, McPherson MA, Dormer R. Metabolism of inositol 1,3,4,5-tetrakisphosphate by human erythrocyte membranes. *Biochem J* 1988;251:927–929.
17. Allan D, Michell RH. A calcium-activated polyphosphoinositide phosphodiesterase in the plasma membrane of human and rabbit erythrocytes. *Biochim Biophys Acta* 1978;508: 277–286.
18. Ellis RB, Galliard T, Hawthorne JN. Phosphoinositides 5. The inositol lipids of ox brain. *Biochem J* 1963;88:125–131.
19. Dangelmaier CA, Daniel JL, Smith JB. Determination of basal and stimulated levels of inositol triphosphate in [^{32}P]orthophosphate-labelled platelets. *Anal Biochem* 1986;154: 414–419.
20. Shayman JA, BeMent DM. The separation of myo-inositol phosphates by ion pair chromatography. *Biochem Biophys Res Commun* 1988;151:114–122.
21. Bidlingmeyer BA, Deming SN, Price WP Jr, Sachok B, Petruoek M. Retention mechanism for reversed-phase ion-pair liquid chromatography. *J Chromatogr* 1979;186:419–434.
22. Mayr GW. A novel metal–dye detection system permits picomolar range h.p.l.c. analysis of inositol polyphosphates from non-radioactively labelled cell or tissue specimens. *Biochem J* 1988;254:585–591.
23. Downes CP, Michell RH. The polyphosphoinositide phosphodiesterase of erythrocyte membranes. *Biochem J* 1981;198:133–140.
24. Lang V, Pryhitka G, Buckley T. Effect of neomycin and ionophore A23187 on ATP levels and turnover of polyphosphoinositides in human erythrocytes. *Can J Biochem* 1977;55:1007–1012.
25. Ponnappa BC, Greenquist AC, Shohet SB. Calcium-induced changes in polyphosphoinositides and phosphatidate in normal erythrocytes, sickle cells and hereditary pyropoikilocytes. *Biochim Biophys Acta* 1980;598:494–501.
26. Allan D, Thomas P. Ca^{2+}-induced biochemical changes in human erythrocytes and their relation to microvesiculation. *Biochem J* 1981;198:433–440.
27. Flatman P, Lew VL. Use of ionophore A23187 to measure and to control free and bound cytoplasmic Mg in intact red cells. *Nature* 1977;267:360–362.
28. Downes CP, Mussat MC, Michell Rh. The inositol trisphosphate phosphomonoesterase of the human erythrocyte membrane. *Biochem J* 1982;203:169–177.
29. Burgess GM, McKinney JS, Irvine RF, Putney JW Jr. Inositol 1,4,5-trisphosphate and inositol 1,3,4-trisphosphate formation in Ca^{2+}-mobilizing hormone-activated cells. *Biochem J* 1985;232:237–243.
30. Burgess GM, Godfrey PP, McKinney JS, Berridge MJ, Irvine RF, Putney JW Jr. The second messenger linking receptor activation to internal Ca release in liver. *Nature* 1984;309: 63–66.
31. Rhoda MD, Sulpice JC, Gascard P, Galacteros F, Giraud F. Endogenous calcium in sickle cells does not activate polyphosphoinositide phospholipase C. *Biochem J* 1988;254: 161–169.

32. Sulpice JC, Gascard P, Journet E, Rendu F, Renard D, Poggioli J. Giraud F. The separation of [^{32}P]inositol phosphates by ion-pair chromatography: optimization of the method and biological applications. *Anal Biochem* 1989;179:90–97.
33. Tranter P, Sulpice JC, Giraud F, Bruckdorfer KR. Inositol phosphate release in human platelets stimulated by low-density lipoproteins. *Biochem Soc Trans* 1989;17:664–665.
34. Poggioli J, Sulpice JC, Vassort G. Inositol phosphate production following α_1-adrenergic, muscarinic or electrical stimulation in isolated rat heart. *FEBS Lett* 1986;206:292–297.
35. Giraud F, Gascard P, Sulpice JC. Stimulation of polyphosphoinositide turnover upon activation of protein kinases in human erythrocytes. *Biochim Biophys Acta* 1988;968: 367–378.
36. Gascard P, Journet E, Sulpice JC, Giraud F. Functional heterogeneity of polyphosphoinositides in human erythrocytes. *Biochem J* 1989;264:547–553.
37. Mauger JP, Poggioli J, Claret M. Synergistic stimulation of the Ca^{2+} influx in rat hepatocytes by glucagon and the Ca^{2+}-linked hormones vasopressin and angiotensin II. *J Biol Chem* 1985;260:11,635–11,642.
38. Levy-Toledano S, Caen JP, Breton-Gorius J, et al. Gray platelet syndrome: α-granule deficiency. *J Lab Clin Med* 1981;98:831–848.

Methods in Inositide Research,
edited by Robin F. Irvine.
Raven Press, Ltd., New York © 1990

6

Nuclear Magnetic Resonance Spectroscopy of *myo*-Inositol Phosphates

Peter Scholz*, Gerhard Bergmann*, and Georg W. Mayr**

**Institute for Analytical Chemistry, Faculty of Chemistry, and **Institute for Physiological Chemistry, Faculty of Medicine, Ruhr-University Bochum, D-4630 Bochum 1, FRG*

The number of inositol phosphate (Ins*P*) isomers identified in animal cells, plants and fungi is steadily increasing. Theoretically, *myo*-inositol alone can exist in 63 different (non-cyclically) phosphorylated derivatives (6 mono-, 15 bis-, 20 tris-, 15 tetrakis-, 6 pentakis- and 1 hexakisphosphate). Identification of new isomers in one of these five differently phosphorylated groups requires sophisticated methods. The earliest developed technique is the chemical analysis by oxidation with periodate, reduction, dephosphorylation, and subsequent identification of polyols found (ORDP technique). Although limited yields of polyols often cause ambiguities, ORDP still is the method of choice when very small amounts of isomers or enantiomers are to be analysed (see Chapter 2). The second most frequently used technique employs isomer-selective high-performance liquid chromatography (HPLC) separation protocols, which are described in several chapters of this volume (see Chapters 2, 5, 7 and 11). But even with optimized separation systems the number of isomers which can be determined unequivocally by HPLC is much less than the theoretical number. Furthermore autologous standards for novel isomers are often unavailable. In the past few years, high-resolution nuclear magnetic resonance (NMR) spectroscopy has evolved as the third useful technique in the identification of novel inositol phosphate isomers, natural inositol phosphate derivatives, and their synthetic analogues. For pure substances (1,2,3,4,5,6,7) or even mixtures of inositol phosphates with phosphocarbohydrates (8), ^{1}H-NMR spectroscopy rapidly provides enough information about the isomeric nature of the inositol phosphate(s) present. Furthermore, the substances are not destroyed by the analysis. With this technique, 39 out of 63 theoretically possible *myo*-inositol phosphate isomers can be identified. Only for the 24 enantiomer pairs among these isomers are absolute configurations indistinguishable by NMR.

In ^{13}C-NMR spectra of inositol phosphates, the six ring-carbon resonances are spread over a wider magnetic field range (between about 69 and 79 p.p.m. downfield of the trimethylsulphoxide (TMS) resonance) than the six ring-proton resonances in corresponding ^{1}H-NMR spectra (between 3.2 and 4.8 p.p.m. downfield of TMS). Both types of resonance are strongly downfield shifted when a phosphate group is located at the same ring position. This behaviour depends strongly on the deuteration of the phosphate groups, i.e., on the pH* value. Nonetheless, this phenomenon facilitates a correct structural assignment. Normally, ^{1}H-NMR spectra of inositol phosphates have several advantages over ^{13}C-NMR spectra. One comes from the higher sensitivity of ^{1}H-NMR spectrometry compared with ^{13}C-NMR spectrometry, which is mainly due to the differing natural abundance of both nuclei (100% versus 1.1%). The other comes from the kind of coupling phenomena the ring-carbons and ring-protons, respectively, are exhibiting. In ^{13}C-NMR spectra, all carbon resonance splittings which lead to overall resonance broadenings between <1 Hz and about 10 Hz are exclusively due to couplings to phosphorus nuclei. Both vicinal spin–spin couplings with phosphorus nuclei at neighbouring ring-carbons (with $^{3}J_{CCOP}$ strongly varying between <1 Hz and about 8 Hz) and geminal coupling to a phosphorus at the same ring-carbon (with a constant $^{2}J_{COP}$ of about 5.4 Hz) can contribute to the overall pattern. When both types of spin–spin coupling occur together, complex and ambiguous patterns can result. The ring-protons, which are the only inositol phosphate protons not exchanging with deuterons (D) in D_2O[1] are held in "fixed" positions by the constant chair conformation of the *myo*-inositol ring. The resulting almost invariable patterns of coupling with neighbouring ring-protons strongly facilitate a positional assignment of these resonances. In addition, these protons exhibit only vicinal coupling to phosphorus nuclei at the same inositol ring position with a relatively constant $^{3}J_{HCOP}$ of about 8.5 Hz (see data below). Because of this, phosphate groups can be assigned unambiguously to individual inositol ring positions in ^{1}H-NMR spectra. In a few instances only, especially when mixtures of inositol phosphates are analysed, a severe crowding of resonances might preclude a structural assignment of all species present from a one-dimensional ^{1}H-NMR spectrum. In such a case, all or most structures can still be assigned when a two-dimensional proton–proton correlation experiment (correlated spectroscopy, COSY) is performed or when a ^{13}C-NMR spectrum is additionally accumulated.

With a sample containing about 1 μmol of the inositol phosphate of interest, the accumulation time required on a 400 MHz instrument is less than 1 h for a one-dimensional ^{1}H-NMR and about 12 h for a ^{13}C-NMR spectrum. When longer accumulation times are available, about a fifth of this amount is the minimum.

[1] Many inositol phosphates, in particular those with higher phosphate content, are insoluble in organic solvents. D_2O is the standard solvent employed for analysis. Exchange of all hydroxyl group and phosphate protons by deuterons dramatically simplifies the ^{1}H-NMR spectra.

Over the past years, 34 different *myo*-inositol phosphate isomers have been isolated or prepared in our laboratory by different techniques (3,8,9,10) and subjected to ^{1}H-NMR and ^{13}C-NMR spectroscopy on a 400 MHz spectrometer using precisely the same pH* values (11) and accumulation conditions for all measurements. For an unambiguous establishment of the database presented below, a number of ^{13}C–^{1}H-COSY experiments have also been performed (data not shown). This experimental database has been used to construct and steadily refine a complete set of spectral data of all NMR-spectroscopically different *myo*-inositol phosphate isomers and to design a computerized search algorithm (INSBASE; P. Scholz and G. W. Mayr, unpublished) as tools for the assignment of correct isomer structures. These tools are valuable for the biochemist identifying naturally occurring inositol phosphate isomers, mixtures, or potential derivatives of the same, as well as for the organic chemist synthesizing *myo*-inositol phosphate isomers or analogues.

PREPARATION OF SAMPLES

The amount of substance should be known prior to the NMR experiment. It is preferable to have more than 1 μmol of the isomer of interest in the NMR tube. Since some inositol phosphates are present in tissues or cells in less than micromolar concentrations, this might require the extraction of kilograms of tissue or cultured cells. Suitable techniques have been described (3,8,10,12) but in particular with cultured cells the preparation of this amount of material might be impossible or extremely tedious. In such a case, about 0.15 μmol of substance provides acceptable ^{1}H-NMR spectra if an accumulation time of up to 12 h is available. For a modern 600 MHz instrument even a 0.1 μmol sample may be adequate. Total phosphate determination after sulphuric acid hydrolysis (8,9,12) provides a reasonable estimate of the amount of substance when the degree of phosphorylation of the inositol phosphate is known. This is normally the case when a substance has been prepared by anion exchange liquid chromatography or analysed by anion exchange HPLC, e.g., the HPLC–metal–dye detection method developed in our laboratory for a direct microanalysis of non-radioactively labelled inositol phosphates (13; and see Chapter 7). As will be pointed out below, knowing the number of phosphate groups of the inositol phosphate(s) under investigation is particularly important for a correct structure assignment and every effort should be undertaken to obtain this figure. Phillippy et al. (10) have successfully performed positive-ion fast-atom bombardment mass spectrometry for this purpose.

Total Phosphorus Determination

Samples containing about 5–10 nmol of total phosphorus are pipetted into new V-shaped glass tubes (type N13-1TK, Macherey & Nagel, Dueren, FRG) and 100 μl of 0.5 M sulphuric acid (p.a. quality) is added. The tubes are kept in an oven at 160°C for 80 min. After that time, 100 μl of P_i-free doubly distilled water is added and the tube kept at 120°C for 20 min. After this treatment, which hydrolyses polyphosphates formed during the first heating period, P_i-free water is again added to make up exactly 100 μl. Portions (20 μl) hereof are subjected to P_i microanalysis, e.g., by the method described by Lanzetta et al. (14). The final sulphuric acid concentration in the assay mixture should be constantly adjusted to (maximally) 0.1 M. P_i standards must be prepared in sulphuric acid of the same concentration. For a correct determination of the organic phosphate, the P_i present in a non-hydrolysed sample has to be subtracted.

Sample Pretreatment for NMR Spectroscopy

Freeze-dried samples are dissolved in about 1 ml of D_2O with an isotopic purity of 99.7 (v/v) (in new 3 ml polypropylene tubes), and treated with about 100 mg (wet weight) of Chelex 100 resin. Prior to use, this resin (from BioRad, Na-form, 100–200 mesh, p.a. quality) is twice regenerated by cycling with 1 M HCl (suprapure) and doubly distilled and Millipore-deionized water in order to remove trace amounts of paramagnetic cations, and finally washed twice and kept suspended in D_2O of 99.7% purity. After about 30 min at ambient temperature with intermittent swirling, the sample solution is carefully removed from the Chelex beads with an Eppendorf pipette and these are washed with another 200 μl of D_2O. The freeze-dried sample (combined with this wash) is once more freeze-dried from 0.5 ml of 99.7% D_2O and once from D_2O of 99.996%. After roughly adjusting the pH* value (11) to 6 or 9 with 2 M DCOOD or 2 M ND_3OD, respectively (by checking 1 μl portions on pH paper), the samples are freeze-dried again. They are finally dissolved in 0.7 ml D_2O (99.996 vol.% D), transferred to a 1.3 ml Eppendorf tube, and adjusted precisely to a pH* value of 6.0 or 9.0, respectively, against a pH-microelectrode (type Ingold 405-M5) which has been equilibrated for 1 h against a saturated solution of KCl in D_2O. About 4 μmol of acetone (0.3 μl, p.a. quality) is added to the sample as a pH*- and HDO-independent internal chemical shift reference. Samples are then stored frozen under liquid nitrogen or transferred directly into 5 mm diameter NMR tubes which have been repeatedly washed with 0.1 M HCl (p.a.), doubly distilled water, and acetone (p.a.), and finally dried in a stream of nitrogen gas (by use of a Teflon capillary inserted into the tube).

NMR Analysis

When trace amounts of paramagnetic cations are left after the above sample treatment, they normally cause a stronger resonance broadening at pH* 9.0

than at pH* 6.0. It is advisable, therefore, to start the NMR examination of a probe with a ^{1}H-NMR spectrum at the latter pH* value. When a crowding of resonances in a resulting spectrum prohibits a structural assignment, a change of the pH* to 9.0 is often sufficient to obtain the desired resolution of these resonances. Only when very complex spectra due to isomeric mixtures or mixtures with phosphorylated sugars are observed at pH* 6.0, a ^{1}H–^{1}H-COSY experiment or, eventually, a ^{13}C-NMR spectrum should first be performed at unchanged pH*. Employing DCOOD and ND_3OD, respectively, for adjusting the pH* values allows for repeated changes between these two pH* values if necessary, since the ND_4COOD formed is removed by freeze-drying.

NMR Instrumentation

Any modern Fourier-transform spectrometer with a proton resonance frequency of 400 MHz or more is adequate. All of our sample spectra have been accumulated using a Bruker AM-400 spectrometer equipped with an ASPECT computer. The resonance frequency is 400.13 MHz for protons and 100.61 MHz for ^{13}C atoms, respectively. Spectrometers with a weaker magnetic field require a larger amount of substance due to the lower sensitivity and lead to more complex spectra. They cannot be recommended.

One-Dimensional ^{1}H-NMR and ^{13}C-NMR Spectra

Both proton and ^{13}C spectra are accumulated at 300 K. For proton spectra, excitation pulses of $\pi/2$ and relaxation delays of 11 to 22 s are employed. The spectral width is set to 4000 Hz at a data size of 32,768 words, which yields a digital resolution of 0.224 Hz/point. The total time requirement for a proton spectrum is between 15 min and 1 h when 1–2 μmol of the inositol phosphate of interest is present. The chemical shifts are given relative to TMS but actually acetone is the internal standard. A chemical shift in D_2O of 2.225 p.p.m. is used for calibration. The carbon spectra are recorded at a resonance frequency of 100.614 MHz, with a shorter excitation pulse and without relaxation delays. The spectral width is 25,000 Hz at the same data size as for proton spectra, which leads to a digital resolution of 1.526 Hz/point. For suppression of the proton coupling a WALTZ 16 decoupling technique is employed which is a modern pulse technique equivalent to the former broadband decoupling. It requires much less decoupling power than the broadband technique and thus reduces temperature instabilities. As for the proton spectra, acetone is used as an internal standard. Its chemical shift of 29.8 p.p.m. is used to evaluate the chemical shift relative to TMS. The accumulation time required for carbon spectra of 1 to 2 μmol of substance is between 4 and 14 h.

^1H–^1H-COSY Experiments

The COSY-45 technique (15) with a pulse sequence of $\pi/2$-τ-$\pi/4$-acq is employed. By this technique, diagonal signals are reduced in intensity. The spectra are recorded between 1.5 and 5.5 p.p.m., with a spectral width of 1602.5 Hz. The experimental parameters for the two-dimensional experiments are set up assuming a ^1H–^1H-coupling constant of 7.5 Hz. For recording, a 256 word × 1024 word data matrix is employed, which is zero-filled prior to Fourier-transformation in order to obtain a 1024 × 2048 spectral data matrix. The resulting resolution is 0.78 Hz/point.

STRUCTURE ASSIGNMENT

Application of the Chemical Shift and Coupling Pattern Database

In order to generate the spectral data of the inositol phosphates not available for an NMR analysis up to now, an algorithm for the prediction of their chemical shifts was developed. Since all substances have the same backbone structure and conformation[2] and have been measured at similar concentrations, same counterion composition (ammonium ions), temperature, and pH* value a simple incremental algorithm could be employed for both types of nuclei:

$$\delta H_i(\text{Ins}P_x) = \delta H_i(\text{Ins}) + \alpha_i a + \beta_i b + \gamma_i c$$

and

$$\delta C_i(\text{Ins}P_x) = \delta C_i(\text{Ins}) + \alpha_i a + \beta_i b + \gamma_i c$$

respectively with:

a = 1 if position i is phosphorylated or 0 if not;
b = 0, 1, or 2 depending on how many nearest neighbours are phosphorylated;
c = 0, 1, 2 depending on how many nearest-but-one neighbours are phosphorylated.

The best fit between measured and calculated chemical shifts was obtained when the axis of symmetry through position 2 and 5 of the *myo*-inositol ring was taken into account. Therefore, the coefficients α, β, γ have been calculated separately for $H_{1/3}$, $H_{4/6}$, H_2, and H_5, and $C_{1/3}$, $C_{4/6}$, C_2, and C_5, respectively. Their p.p.m. values are given in Table 1. Tables 2 to 4 comprise the chemical shifts observed for substances subjected to ^1H- and ^{13}C-

[2] From all *myo*-inositol phosphate spectra measured up to now over a pH* range between 4 and 11 there is no evidence for any other than the chair conformation in which OH_2 is oriented perpendicular to the plane through carbons 1, 3, 4, and 6.

TABLE 1. *Incremental parameters for chemical shift algorithms*

Parameter	pH* 6.0				pH* 9.0			
	$H_{1/3}$	$H_{4/6}$	H_2	H_5	$H_{1/3}$	$H_{4/6}$	H_2	H_5
α	+0.426	+0.504	+0.477	+0.523	+0.361	+0.435	+0.344	+0.446
β	+0.069	+0.131	+0.169	+0.154	+0.041	+0.111	+0.161	+0.152
γ	+0.054	+0.065	+0.032	+0.067	+0.044	+0.084	+0.050	+0.050
	$C_{1/3}$	$C_{4/6}$	C_2	C_5	$C_{1/3}$	$C_{4/6}$	C_2	C_5
α	+3.504	+4.860	+3.988	+4.432	+2.480	+3.396	+3.957	+3.256
β	−0.309	−0.641	−0.780	−0.484	+0.075	−0.313	−0.023	−0.185
γ	−0.070	+0.178	−0.013	−0.100	−0.041	+0.462	−0.223	+0.205

A least-squares technique was employed for estimating the sets of chemical shift increments (α, β, γ, given in p.p.m. downfield) for each pH* value and each of the four groups of resonances $H_{1/3}$, $H_{4/6}$, H_2, and H_5 employing the equations described in the text. With these parameters, the chemical shifts were recalculated. Average deviations (± S.E., cases in parentheses) between calculated and observed chemical shift values of a set of spectra were (in p.p.m.): 0.032 ± 0.026 (n = 210) and 0.040 ± 0.030 (n = 168) for ^1H resonances at pH* 6.0 and 9.0, respectively; 0.235 ± 0.180 (n = 168) and 0.334 ± 0.243 (n = 114) for ^{13}C resonances at pH* 6.0 and 9.0, respectively. Chemical shifts of ^{13}C resonances at pH* 9.0 can be calculated from the corresponding incremental parameters and the *myo*-inositol chemical shifts.

NMR spectroscopy at pH* 6.0 and to ^1H-NMR analysis at pH* 9.0 and the chemical shifts calculated by this algorithm for substances not measured. Each of the three sets of data in Tables 2, 3, 4 comprises the 29 one- to six-fold phosphorylated *myo*-inositol phosphate isomers which can be discriminated by NMR spectroscopy[3] (16) and the parent compound, *myo*-inositol.

A computer-aided search algorithm strongly facilitates structural assignments of measured inositol phosphate spectra by matching some or all of their resonances with corresponding ones in the database. Three selection criteria are employed. Their efficiency in the search process has been evaluated precisely via the probability of refinding a known spectrum in the database. For counting the probability of correct matches, each spectrum was removed individually from the database, and chemical shifts for unknown spectra, including the removed one, were recalculated from the diminished experimental database. Finally, the removed spectrum was searched again in the database. The procedure was repeated for all combinations of one to six resonances.

In the search process, the sum-of-squares of errors between the resonances of a measured spectrum serves as a starting criterion and provides a rank order of matching spectra. Figures 1a to 1c show that with this

[3] The D- and L-configuration of a pair of enantiomers of a *myo*-inositol phosphate isomer cannot be discriminated by NMR spectroscopy, since both configurations are magnetically equivalent.

TABLE 2. Proton resonances of *myo*-inositol phosphates at pH* 6.0

Substance	C/M[a]	H_1		H_2		H_3		H_4		H_5		H_6	
myo-Inositol	M	3.52	dD	4.04	t	3.52	dD	3.61	T	3.26	T	3.61	T
Ins(1)P	M	3.92	dT	4.24	t	3.56	dD	3.64	T	3.33	T	3.74	T
Ins(2)P	M	3.50	dD	4.51	tD	3.50	dD	3.71	T	3.25	T	3.71	T
Ins(4)P	M	3.55	dD	4.06	t	3.65	dD	4.13	Q	3.43	T	3.69	T
Ins(5)P	C	3.57	dD	4.04	t	3.57	dD	3.74	T	3.78	Q	3.74	T
Ins(1,2)P_2	M	3.98	dT	4.66	tD	3.49	dD	3.74	T	3.33	T	3.83	T
Ins(1,3)P_2	M	3.98	dT	4.39	t	3.98	dT	3.78	T	3.41	T	3.78	T
Ins(1,4)P_2	M	3.96	dT	4.25	t	3.69	dD	4.16	Q	3.49	T	3.82	T
Ins(1,5)P_2	M	3.96	dT	4.26	t	3.63	dD	3.80	T	3.78	Q	3.82	T
Ins(1,6)P_2	M	3.98	dT	4.25	t	3.59	dD	3.71	T	3.48	T	4.21	Q
Ins(2,4)P_2	M	3.54	dD	4.56	tD	3.68	dD	4.21	Q	3.46	T	3.77	T
Ins(2,5)P_2	C	3.64	dD	4.51	tD	3.64	dD	3.80	T	3.78	Q	3.80	T
Ins(4,5)P_2	M	3.62	dD	4.06	t	3.70	dD	4.23	Q	3.95	Q	3.82	T
Ins(4,6)P_2	C	3.58	dD	4.10	t	3.58	dD	4.17	Q	3.56	T	4.17	Q
Ins(1,2,3)P_3	M	4.02	dT	4.72	tD	4.02	dT	3.85	T	3.50	T	3.85	T
Ins(1,2,4)P_3	M	4.02	dT	4.73	tD	3.70	dD	4.23	Q	3.53	T	3.87	T
Ins(1,2,5)P_3	M	4.07	dT	4.72	tD	3.62	dD	3.88	T	3.89	Q	3.94	T
Ins(1,2,6)P_3	M	4.10	dT	4.70	tD	3.56	dD	3.79	T	3.50	T	4.30	Q
Ins(1,3,4)P_3	M	4.01	dT	4.39	t	4.08	dT	4.31	Q	3.55	T	3.87	T
Ins(1,3,5)P_3	M	4.06	dT	4.47	t	4.06	dT	3.93	T	3.92	Q	3.93	T
Ins(1,4,5)P_3	M	4.03	dT	4.29	t	3.74	dD	4.27	Q	4.02	Q	3.93	T
Ins(1,4,6)P_3	C	4.01	dT	4.27	t	3.64	dD	4.17	Q	3.63	T	4.30	Q
Ins(1,5,6)P_3	M	4.09	dT	4.22	t	3.67	dD	3.82	T	4.00	Q	4.39	Q
Ins(2,4,5)P_3	M	3.60	dD	4.55	tD	3.72	dD	4.27	Q	3.97	Q	3.88	T
Ins(2,4,6)P_3	M	3.71	dD	4.57	tD	3.71	dD	4.25	Q	3.60	T	4.25	Q
Ins(4,5,6)P_3	M	3.72	dD	4.10	t	3.72	dD	4.32	Q	4.11	Q	4.32	Q
Ins(1,2,3,4)P_4	M	4.05	dT	4.88	tD	4.14	dT	4.30	Q	3.58	T	3.87	T
Ins(1,2,3,5)P_4	C	4.12	dT	4.85	tD	4.12	dT	3.93	T	3.91	Q	3.93	T
Ins(1,2,4,5)P_4	M	4.08	dT	4.78	tD	3.73	dD	4.32	Q	4.04	Q	3.95	T
Ins(1,2,4,6)P_4	M	4.14	dT	4.74	tD	3.74	dD	4.28	Q	3.65	T	4.36	Q
Ins(1,2,5,6)P_4	M	4.18	dT	4.75	tD	3.67	dD	3.90	T	4.04	Q	4.44	Q
Ins(1,3,4,5)P_4	M	4.09	dT	4.43	t	4.15	dT	4.44	Q	4.08	Q	3.94	T
Ins(1,3,4,6)P_4	M	4.11	dT	4.35	t	4.11	dT	4.35	Q	3.65	T	4.35	Q
Ins(1,4,5,6)P_4	M	4.14	dT	4.27	t	3.78	dD	4.35	Q	4.15	Q	4.46	Q
Ins(2,4,5,6)P_4	M	3.74	dD	4.62	tD	3.74	dD	4.38	Q	4.13	Q	4.38	Q
Ins(1,2,3,4,5)P_5	M	4.10	dT	4.90	tD	4.16	dT	4.42	Q	4.08	Q	3.95	T
Ins(1,2,3,4,6)P_5	M	4.13	dT	4.90	tD	4.13	dT	4.34	Q	3.67	T	4.34	Q
Ins(1,2,4,5,6)P_5	M	4.12	dT	4.75	tD	3.70	dD	4.33	Q	4.14	Q	4.42	Q
Ins(1,3,4,5,6)P_5	M	4.13	dT	4.45	t	4.13	dT	4.43	Q	4.14	Q	4.43	Q
Ins(1,2,3,4,5,6)P_6	M	4.15	dT	4.95	tD	4.15	dT	4.42	Q	4.17	Q	4.42	Q

For each proton resonance the chemical shift (in p.p.m. downfield of TMS) and the coupling pattern (denoted by a one- or two-letter symbol) are given. Denoted patterns are depicted in Fig. 2. Subtypes observed for some patterns (see Fig. 2) are indicative of certain conformations. As their discrimination is unimportant for isomer assignments, they are not indicated here. Ranges of underlying coupling constants are given in the legend to Fig. 2.

Abbreviations: Ins$P$, inositol monophosphate; InsP_2, inositol bisphosphate; InsP_3, inositol trisphosphate; InsP_4, inositol tetrakisphosphate; InsP_5, inositol pentakisphosphate; InsP_6, inositol hexakisphosphate. The isomer numbers in each case are indicated by the numbers in parentheses. See also note (3), p. 67.

[a] C and M indicate whether chemical shift data have been calculated from the experimental database or measured, respectively.

TABLE 3. Proton resonances of *myo*-inositol phosphates at pH* 9.0

Substance	C/M[a]	H_1		H_2		H_3		H_4		H_5		H_6	
myo-Inositol	M	3.52	dD	4.04	t	3.52	dD	3.61	T	3.26	T	3.61	T
Ins(1)*P*	M	3.90	dT	4.23	t	3.58	dD	3.64	T	3.34	T	3.75	T
Ins(2)*P*	M	3.47	dD	4.49	tD	3.47	dD	3.73	T	3.25	T	3.73	T
Ins(4)*P*	M	3.56	dD	4.06	t	3.63	dD	4.10	Q	3.42	T	3.70	T
Ins(5)*P*	C	3.56	dD	4.04	t	3.56	dD	3.72	T	3.70	Q	3.72	T
Ins(1,2)P_2	M	3.97	dT	4.54	tD	3.45	dD	3.75	T	3.31	T	3.84	T
Ins(1,3)P_2	M	3.92	dT	4.28	t	3.92	dT	3.75	T	3.38	T	3.75	T
Ins(1,4)P_2	M	3.93	dT	4.22	t	3.67	dD	4.12	Q	3.47	T	3.82	T
Ins(1,5)P_2	M	3.90	dT	4.19	t	3.60	dD	3.75	T	3.75	Q	3.78	T
Ins(1,6)P_2	M	3.90	dT	4.36	t	3.58	dD	3.69	T	3.49	T	4.12	Q
Ins(2,4)P_2	M	3.47	dD	4.51	tD	3.62	dD	4.17	Q	3.45	T	3.81	T
Ins(2,5)P_2	C	3.60	dD	4.38	tD	3.60	dD	3.80	T	3.70	Q	3.80	T
Ins(4,5)P_2	M	3.60	dD	4.02	t	3.70	dD	4.13	Q	3.86	Q	3.82	T
Ins(4,6)P_2	C	3.56	dD	4.13	t	3.56	dD	4.12	Q	3.56	T	4.12	Q
Ins(1,2,3)P_3	C	3.96	dT	4.70	tD	3.96	dT	3.80	T	3.36	T	3.80	T
Ins(1,2,4)P_3	C	3.92	dT	4.59	tD	3.64	dD	4.12	Q	3.46	T	3.88	T
Ins(1,2,5)P_3	C	3.96	dT	4.54	tD	3.64	dD	3.80	T	3.75	Q	3.91	T
Ins(1,2,6)P_3	M	3.95	dT	4.65	tD	3.49	dD	3.81	T	3.51	T	4.18	Q
Ins(1,3,4)P_3	C	3.92	dT	4.41	t	3.96	dT	4.15	Q	3.51	T	3.80	T
Ins(1,3,5)P_3	C	3.96	dT	4.36	t	3.96	dT	3.83	T	3.80	Q	3.83	T
Ins(1,4,5)P_3	M	3.92	dT	4.35	t	3.73	dD	4.15	Q	3.92	Q	3.86	T
Ins(1,4,6)P_3	C	3.92	dT	4.30	t	3.60	dD	4.12	Q	3.61	T	4.23	Q
Ins(1,5,6)P_3	M	3.98	dT	4.32	t	3.63	dD	3.83	T	3.88	Q	4.33	Q
Ins(2,4,5)P_3	M	3.49	dD	4.46	tD	3.73	dD	4.18	Q	3.83	Q	3.93	T
Ins(2,4,6)P_3	M	3.62	dD	4.49	tD	3.62	dD	4.20	Q	3.53	T	4.20	Q
Ins(4,5,6)P_3	M	3.71	dD	4.04	t	3.71	dD	4.25	Q	4.06	Q	4.25	Q
Ins(1,2,3,4)P_4	M	3.98	dT	4.69	tD	3.98	dT	4.24	Q	3.56	T	3.91	T
Ins(1,2,3,5)P_4	C	4.00	dT	4.70	tD	4.01	dT	3.91	T	3.80	Q	3.91	T
Ins(1,2,4,5)P_4	M	4.01	dT	4.70	tD	3.68	dD	4.32	Q	3.97	Q	4.01	T
Ins(1,2,4,6)P_4	M	3.99	dT	4.63	tD	3.60	dD	4.22	Q	3.58	T	4.29	Q
Ins(1,2,5,6)P_4	M	4.06	dT	4.59	tD	3.53	dD	3.93	T	3.87	Q	4.37	Q
Ins(1,3,4,5)P_4	M	3.95	dT	4.53	t	4.01	dT	4.34	Q	3.93	Q	3.91	T
Ins(1,3,4,6)P_4	C	3.96	dT	4.46	t	3.96	dT	4.23	Q	3.66	T	4.23	Q
Ins(1,4,5,6)P_4	M	3.97	dT	4.39	t	3.75	dD	4.21	Q	4.06	Q	4.40	Q
Ins(2,4,5,6)P_4	M	3.66	dD	4.50	tD	3.66	dD	4.29	Q	4.06	Q	4.29	Q
Ins(1,2,3,4,5)P_5	M	3.98	dT	4.71	tD	4.00	dT	4.41	Q	3.97	Q	3.95	T
Ins(1,2,3,4,6)P_5	M	3.98	dT	4.69	tD	3.98	dT	4.32	Q	3.62	T	4.32	Q
Ins(1,2,4,5,6)P_5	M	3.99	dT	4.57	tD	3.67	dD	4.24	Q	4.03	Q	4.36	Q
Ins(1,3,4,5,6)P_5	M	3.97	dT	4.51	t	3.97	dT	4.30	Q	4.03	Q	4.30	Q
Ins(1,2,3,4,5,6)P_6	M	4.10	dT	4.75	tD	4.10	dT	4.47	Q	4.17	Q	4.47	Q

See note to Table 2 for further explanations.

[a] C and M indicate whether chemical shift data have been calculated from the experimental database or measured, respectively.

criterion alone it is possible to achieve assignments that are about 90% correct, if all six resonances can be read from a ^{1}H- or ^{13}C-NMR spectrum.

For proton resonances of *myo*-inositol phosphates, the vicinal ^{1}H–^{1}H- and ^{1}H–^{31}P coupling patterns can be reliably predicted (see Introduction, above). All coupling patterns possible for protons $H_{1/3}$, H_2, $H_{4/6}$ and H_5 are depicted in Fig. 2. Ranges of corresponding coupling constants are given in the subscript to this figure. The symbolic abbreviations given to these basic coupling

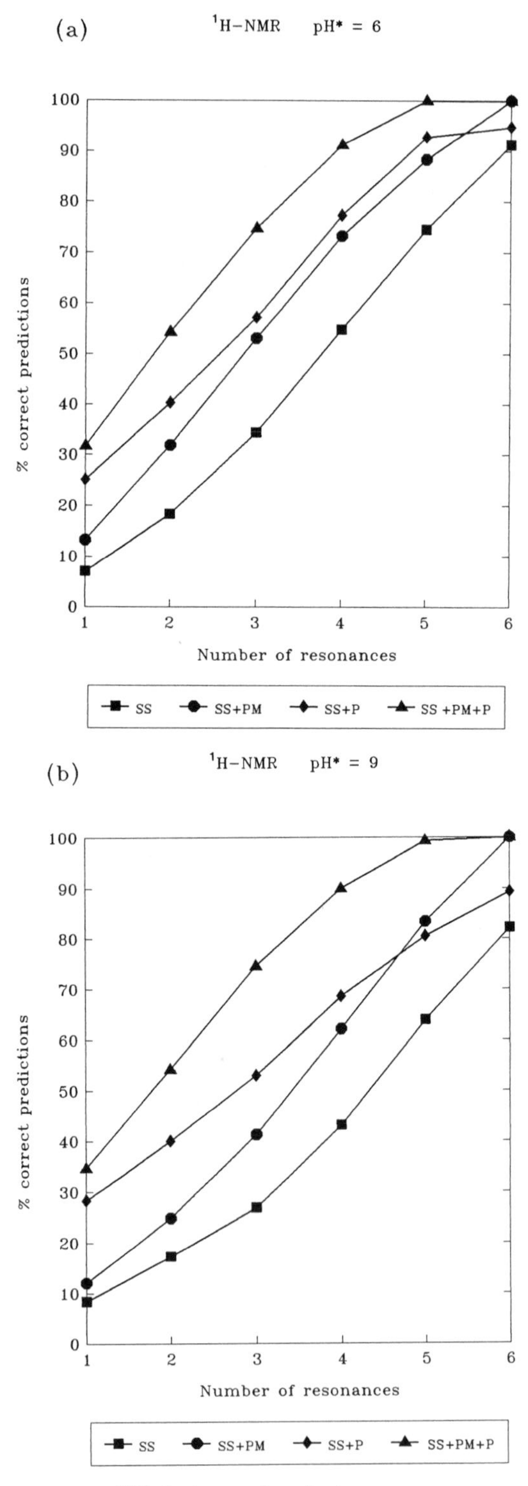

FIG. 1. Legend on facing page.

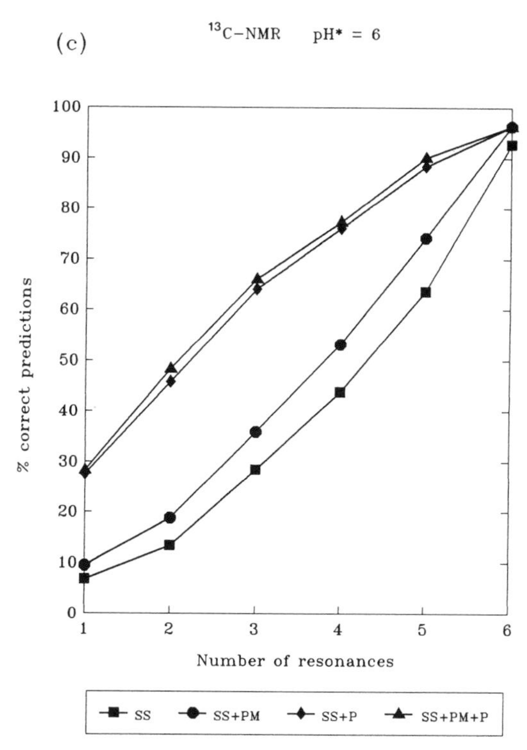

FIG. 1. Probability of correct structural assignments of *myo*-inositol phosphate isomers. Dependence on the number of interpreted resonances and on search criteria. The data treatment resulting in the percentage probabilities depicted is explained in the text. **(a)** Probability of correct structure assignments from one to six proton resonances measured at pH* 6.0. Search criteria employed are: squares (SS), sum of squares alone; dots (SS + PM), sum of squares plus matching of coupling patterns; diamonds (SS + P), sum of squares plus correct degree of phosphorylation; triangles (SS + PM + P), all three criteria. **(b)** Probability of correct structure assignments from one to six proton resonances measured at pH* 9.0. Search criteria and symbols denoting them as in **(a)**. **(c)** Probability of correct structure asignments from one to six ^{13}C resonances measured at pH* 9.0. Search criteria and symbols denoting them are as in **(a)**.

patterns are listed in Tables 2 and 3, together with the chemical shift values. Their matching with an observed set of resonances allows an unequivocal assignment to an isomer unless some observed resonances are severely crowded. Incorporating a matching of coupling patterns in the search improves the probability of correct structure assignments to 100% and 80% in the case of six and five readable proton resonances, respectively (Figs. 1a and 1b).

In the ^{13}C-NMR spectra, the observed resonance splittings can result from both vicinal ($^3J_{CCOP}$ varying between <1 Hz and 8 Hz) and geminal ($^3J_{COP}$ constantly about 5.4 Hz) couplings. The vicinal couplings are strongly de-

TABLE 4. ^{13}C resonances of *myo*-inositol phosphates at pH* 6.0

Substance	C/M[a]	C_1		C_2		C_3		C_4		C_5		C_6	
myo-Inositol	M	70.65	s	71.70	s	70.65	s	71.92	s	73.87	s	71.92	s
Ins(1)P	M	74.69	d	70.97	sd	70.34	s	71.85	s	73.70	s	71.38	sd
Ins(2)P	M	70.97	sd	75.66	d	70.97	sd	72.48	s	74.04	s	72.48	s
Ins(4)P	M	70.55	s	71.43	s	70.64	sd	76.60	d	73.46	sd	71.77	s
Ins(5)P	C	70.58	s	71.70	s	70.58	s	71.28	sd	78.30	d	71.28	sd
Ins(1,2)P_2	M	74.08	dc	74.30	dc	70.73	sd	72.39	s	73.60	s	71.44	sd
Ins(1,3)P_2	C	74.08	d	70.15	dc	74.08	d	71.28	sd	73.67	s	71.28	sd
Ins(1,4)P_2	M	74.73	d	70.62	sd	70.29	sd	76.55	d	73.23	sd	71.13	sd
Ins(1,5)P_2	C	74.08	d	70.93	sd	70.51	s	71.28	sd	78.20	d	70.64	dc
Ins(1,6)P_2	C	73.84	dc	70.91	sd	70.58	s	72.09	s	73.29	sd	76.13	dc
Ins(2,4)P_2	M	70.71	sd	75.98	d	70.17	dc	77.22	d	73.71	sd	72.22	s
Ins(2,5)P_2	M	70.61	sd	75.34	d	70.61	sd	72.04	sd	78.32	d	72.04	sd
Ins(4,5)P_2	M	70.26	s	71.16	s	70.71	sd	75.92	dc	77.84	dc	71.68	sd
Ins(4,6)P_2	C	70.35	sd	71.68	s	70.35	sd	76.95	d	72.91	dc	76.95	d
Ins(1,2,3)P_3	C	73.77	dc	74.12	c	73.77	dc	71.46	sd	73.67	s	71.46	sd
Ins(1,2,4)P_3	C	73.84	dc	74.89	dc	69.97	dc	76.95	d	73.29	sd	71.64	sd
Ins(1,2,5)P_3	M	73.91	dc	74.34	dc	70.25	sd	71.62	sd	78.36	d	70.65	dc
Ins(1,2,6)P_3	M	73.14	c	74.51	dc	70.64	sd	72.09	s	73.54	sd	76.48	dc
Ins(1,3,4)P_3	C	74.08	d	70.13	dc	73.77	dc	76.13	dc	73.19	sd	71.46	sd
Ins(1,3,5)P_3	M	74.28	d	69.67	dc	74.28	d	70.61	dc	78.14	d	70.61	dc
Ins(1,4,5)P_3	M	74.34	d	70.28	sd	70.38	sd	75.79	dc	77.69	dc	71.04	dc
Ins(1,4,6)P_3	C	73.84	dc	70.90	sd	70.28	sd	76.95	d	72.81	dc	76.31	dc
Ins(1,5,6)P_3	M	74.33	dc	71.01	sd	69.85	s	71.28	sd	78.06	dc	75.72	c
Ins(2,4,5)P_3	M	70.52	sd	75.47	d	70.26	dc	76.48	dc	77.90	dc	72.25	sd
Ins(2,4,6)P_3	M	70.10	dc	75.87	d	70.10	dc	77.04	d	73.11	dc	77.04	d
Ins(4,5,6)P_3	M	70.23	sd	70.90	s	70.23	sd	76.48	dc	77.10	c	76.48	dc
Ins(1,2,3,4)P_4	M	73.95	dc	74.40	c	72.95	c	76.33	dc	72.93	sd	70.76	sd
Ins(1,2,3,5)P_4	C	73.70	dc	74.12	c	73.70	dc	70.82	dc	78.10	d	70.82	dc
Ins(1,2,4,5)P_4	M	73.86	dc	74.60	dc	69.88	dc	76.42	dc	77.62	dc	70.96	dc
Ins(1,2,4,6)P_4	M	73.10	c	75.48	dc	69.89	dc	76.95	d	72.88	dc	76.56	dc
Ins(1,2,5,6)P_4	M	73.25	c	75.10	dc	70.25	sd	71.48	sd	78.10	dc	75.82	c
Ins(1,3,4,5)P_4	M	73.91	d	70.13	dc	74.09	dc	75.60	c	77.78	dc	70.55	dc
Ins(1,3,4,6)P_4	C	73.77	dc	70.12	dc	73.77	dc	76.31	dc	72.71	dc	76.31	dc
Ins(1,4,5,6)P_4	M	74.16	dc	70.80	sd	69.87	sd	76.30	dc	77.04	c	75.82	c
Ins(2,4,5,6)P_4	M	69.98	dc	75.61	d	69.98	dc	76.87	dc	77.19	c	76.87	dc
Ins(1,2,3,4,5)P_5	M	73.69	dc	74.20	c	73.01	c	75.45	c	77.51	dc	70.65	dc
Ins(1,2,3,4,6)P_5	M	72.94	c	75.19	c	72.94	c	76.11	dc	72.84	dc	76.11	dc
Ins(1,2,4,5,6)P_5	M	73.11	c	74.60	dc	69.94	dc	76.75	dc	76.98	c	75.83	c
Ins(1,3,4,5,6)P_5	M	73.51	dc	70.22	dc	73.51	dc	75.53	c	76.98	c	75.53	c
Ins(1,2,3,4,5,6)P_6	C	73.39	c	74.10	c	73.39	c	75.85	c	77.13	c	75.85	c

For each ^{13}C resonance the chemical shift (in p.p.m. downfield of TMS) and potential coupling patterns (denoted by one- or two-letter symbols) are given. Denoted patterns are depicted in Fig. 3; corresponding coupling constants are given in the legend to Fig. 3. Since subtypes c_1, c_2 and c_3 are hard to discriminate in many cases, all resonances more complex than doublets are assigned to one complex group, denoted c.

[a] C and M indicate whether chemical shift data have been calculated from the experimental database or measured, respectively.

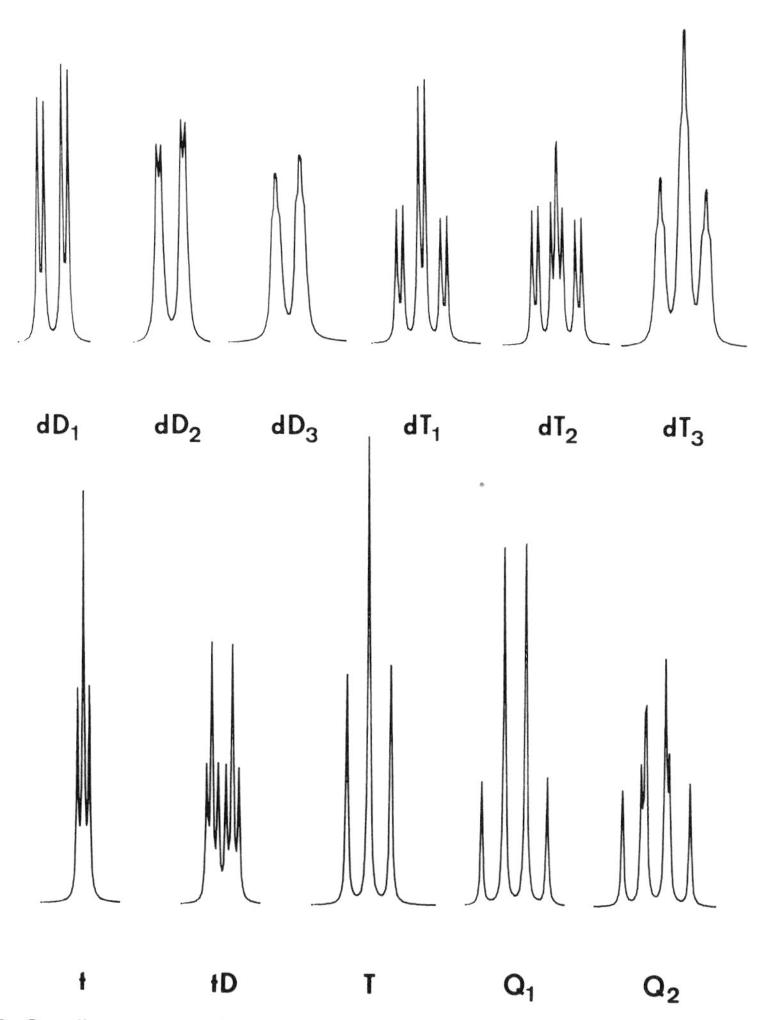

FIG. 2. Coupling patterns of proton resonances of *myo*-inositol phosphates. In the upper row, coupling patterns of protons 1 and 3, respectively, are depicted. When the corresponding ring positions are not phosphorylated, patterns dD_1 or dD_2 are normally observed. Underlying coupling constants are (in means ± S.D. (range)): $^3J_{HCCH,cis}$ 2.39 ± 0.46 (1.3–2.9); $^3J_{HCCH,trans}$ 9.48 ± 0.38 (8–10.3). When the ring positions are phosphorylated, the additional HCOP coupling leads to patterns dT_1 or dT_2 ($^3J_{HCOP}$ 8.62 ± 0.93 (6.4–10.5)). Only when position 2 is phosphorylated, does a further splitting of the resonance (long-range coupling with phosphate 2, $^4J_{HCCOP}$ 1.12 ± 0.32 (0–2.34)) may occur. Patterns dD_3 and dT_3 thus are indicative of phosphate at position 2. The left two patterns in the lower row are characteristic for H_2. When this position of the ring is phosphorylated, a splitting of t (see $^3J_{HCCH,cis}$, above) into tD ($^3J_{HCOP}$ 8.26 ± 1.18 (5.9–10.3)) is the consequence. Pattern T (see $^3J_{HCCH,trans}$, above) is characteristic for resonances H_4, H_5 and H_6, respectively; phosphorylation at the corresponding ring position leads to patterns Q_1 or Q_2, respectively ($^3J_{HCOP}$ 8.62 ± 0.93 (6.4–10.5)).

pendent on the rotation of the neighboring phosphate groups around the C–O axis, i.e., the dihedral angle formed between the C—C and O—P bonds of the corresponding CCOP structure. Because of this variability, only three groups of coupling patterns are discriminated. Possible variants within each of these groups have been simulated and are depicted in Fig. 3. Again the symbols denoting the depicted groups of coupling patterns are given in the corresponding database (Table 4) together with the chemical shifts. Although these complexities render a matching of coupling patterns a weak selection criterion for ^{13}C-NMR spectra (compare the corresponding curve in Fig. 1C with those in Figs. 1a and 1b) it still rules out about 10% of assignments as being incorrect.

Whenever the degree of phosphorylation of an inositol phosphate or a group of isomers is known with some certainty, it should also be incorporated as a selection criterion. In this case, 100% of the predictions are correct when six or five proton resonances are readable and 90% if four resonances are employed. For ^{13}C-NMR spectra, this criterion is particularly powerful (see Fig. 1c), since it compensates for the weak pattern matching criterion. The best predictions always are obtained when all three selection criteria are employed together.

The ease of computer-aided searches in the database facilitates correct predictions of inositol phosphate structures even in complex mixtures. UNIX/XENIX and MS-DOS versions of the program INSBASE (P. Scholz and G. W. Mayr, unpublished) and details about its usage are available upon request.

But even without this computer program correct structure assignments are strongly facilitated when the chemical shift data (Tables 2, 3, 4) and the coupling patterns, depicted in Figs. 2 and 3, and tabulated together with the chemical shifts, are carefully checked and when the above discussed rules for assignment are considered.

Protocol for Assigning Spectra from Inositol Phosphate Mixtures

This protocol gives some further working rules for isomer assignments in spectra obtained from mixtures of inositol phosphates (and other phosphorylated carbohydrates). Starting from the resonances with the largest integrals (integrals are much more reliable in ^{1}H-NMR spectra than in ^{13}C-NMR spectra), an assignment of the coupling patterns to groups of nuclei is performed as discussed above. Coupling constants should be checked and must be in the range given in the legends to Figs. 2 and 3. In order to find out whether a mixture of inositol phosphates with other phosphosugars is present (most likely in extracted inositol mono- and bisphosphate fractions) the spectrum should be carefully inspected for anomeric proton or carbon resonances, which in most cases are shifted further downfield than any of the

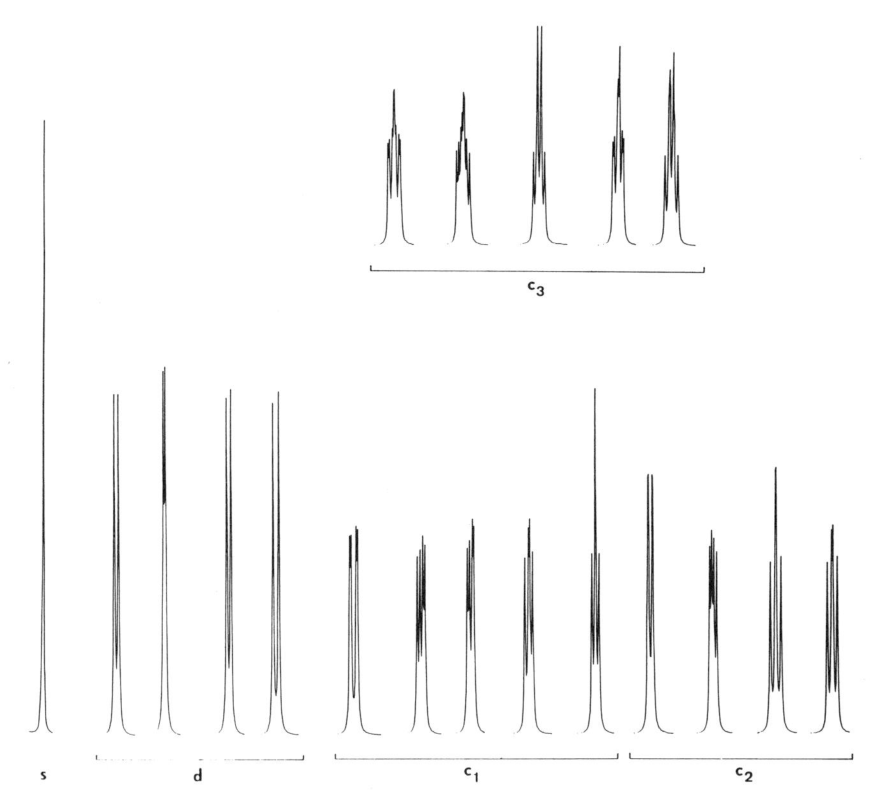

FIG. 3. Coupling patterns of ^{13}C resonances of *myo*-inositol phosphates. A singlet (s) is always observed if a certain ring-carbon is not phosphorylated and not neighboured by a phosphate. Correspondingly, doublets (d) can indicate a geminal coupling to a phosphorus atom at the same position (left doublet, $^{2}J_{COP}$ 5.43 ± 0.41 (4.30–6.28)) or a geminal one to a phosphate group at one neighbouring ring-carbon ($^{3}J_{CCOP}$ 4.48 ± 2.12 (0–8.3)), doublets with 2, 5 and 7 Hz are depicted (from left to right). With $^{3}J_{CCOP}$ becoming <1 Hz, the resonance may appear like a singlet. Complex coupling patterns can result from two neighbouring phosphate groups without a geminal one (c_1, patterns depicted from left to right for ^{3}J values of 8 and 2, 7 and 3, 6 and 2, 6 and 4, and 4 and 4 Hz, respectively), from one geminal and one vicinal coupling, i.e., phosphate at the same and one neighbouring position (c_2, patterns depicted from left to right for ^{2}J of 5.5 Hz and ^{3}J values of 1, 3, 5 and 7 Hz, respectively), and, finally, from one geminal and two vicinal couplings (c_3, patterns depicted in the upper row from left to right for ^{2}J of 5.5 Hz and ^{3}J values of 8 and 2, 7 and 3, 4.5 and 4.5, 6 and 2, and 6 and 4 Hz, respectively). Note that with these complex couplings the overall resonance width can exceed 10 Hz and the amplitudes are becoming very small. When such a complex coupling pattern is not clearly resolved, an inspection of the amplitude and overall width of the resonance normally allows a discrimination from a simple doublet.

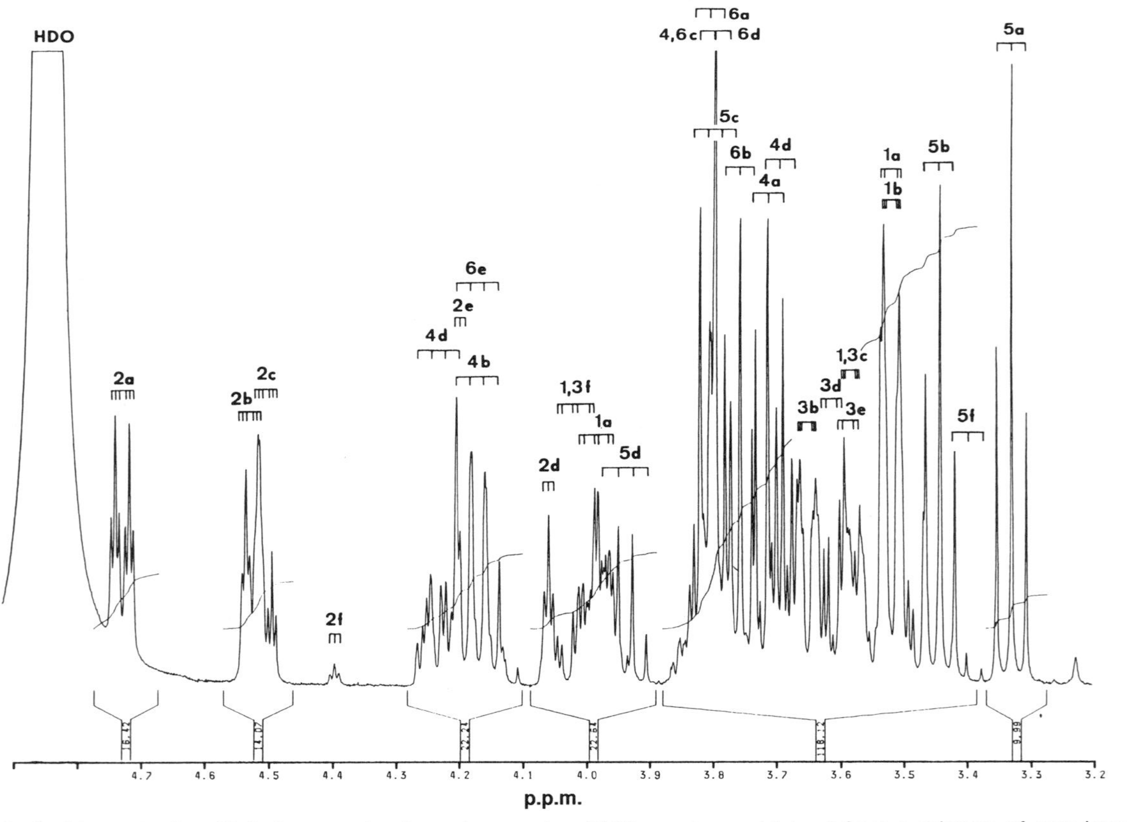

FIG. 4. Assignment of multiple isomer structures in a proton NMR spectrum obtained from a mixture of *myo*-inositol bisphosphates. The total amount of inositol bisphosphates was 15 μmol. Assigned resonances are labelled by the ring-proton number and a letter denoting the identified isomer: a, Ins(1,2)P_2 (~25% of total); b, Ins(2,4)P_2 (~25%); c, Ins(2,5)P_2 (~14%); d, Ins(4,5)P_2 (~11%); e, Ins(1,6)P_2 (~6%); f, Ins(1,3)P_2 (~4%). For abbreviations, see notes to Table 2.

inositol phosphate resonances and in proton spectra usually resonate downfield of HDO (17).

After a successful structural assignment of the dominating set of resonances, the whole procedure is repeated with those remaining resonances (the original resonances minus *all* resonances belonging to the identified isomer) which again exhibit the largest integrals, and so on. When more than six inositol phosphate resonances of similar size are encountered, two equally abundant isomers may exist. In this case, combinations of resonances potentially belonging to one isomer (no more than one H_2, two $H_{1/3}$, two $H_{4/6}$ and one H_5 resonances can be in one set) have to be iteratively tested. From the integrals of non-superposed resonances of assigned inositol phosphate isomers (this is frequently the case for H_2, H_5 and $H_{4/6}$ resonances) their mass relations can be evaluated. An example of a spectrum of a mixture of inositol bisphosphates in which six isomers have been assigned is shown in Fig. 4. Further examples of inositol phosphate structure assignments in complex mixtures have been published by Radenberg et al. (8).

Acknowledgments A number of inositol phosphates subjected to NMR spectroscopy have been kindly provided by Dipl. Chem. Burckhard Koppitz, Dipl. Biol. Thomas Radenberg (both in G. W. M.'s laboratory), and by Dr Barbara Goldschmidt (Perstorp SA, Sweden). The expert technical assistence by Friedhelm Vogel and Cornelia Tietz is highly acknowledged. Thanks are expressed to Dr W. Dietrich for assistance in some of the NMR measurements. This work was supported by grants from the Deutsche Forschungsgemeinschaft (Ma 989), from the Bundesminister für Forschung und Technologie, and from the Fonds der Chemischen Industrie.

REFERENCES

1. Lindon JC, Baker DJ, Farrant RD, Williams JM. ^{1}H,^{13}C and ^{31}P n.m.r. spectra and molecular conformation of *myo*-inositol 1,4,5-trisphosphate. *Biochem J* 1986;233:275–277.
2. Cerdan S, Hansen CA, Johanson R, Inubushi T, Williamson JR. Nuclear magnetic resonance spectroscopic analysis of *myo*-inositol phosphates including inositol 1,3,4,5-tetrakisphosphate. *J Biol Chem* 1986;261:14,767–14,680.
3. Mayr GW, Dietrick W. The only inositol tetrakisphosphate detectable in avian erythrocytes is the isomer lacking phosphate at position 3: a NMR study. *FEBS Lett* 1987;213:278–282.
4. Henne V, Mayr GW, Grabowski B, Koppitz B, Söling HD. Semisynthetic derivatives of inositol 1,4,5-trisphosphate substituted at the 1 phosphate group. *Eur J Biochem* 1988; 174:95–101.
5. Walker JW, Feeney J, Trentham DR. Photolabile precursors of inositol phosphates preparation and properties of 1-(2-nitrophenyl)ethyl esters of *myo*-inositol 1,4,5-trisphosphate. *Biochemistry* 1989;28:3272–3280.
6. Phillippy BQ. Identification by two dimensional NMR of *myo*-inositol tris- and tetrakis(phosphates) formed from phytic acid by wheat phytase. *J Agric Food Chem* 1989;37:1261–1265.
7. Tegge W, Ballou CE. Chiral synthesis of D- and L-*myo*-inositol 1,4,5-trisphosphate. *Proc Natl Acad Sci USA* 1989;86:94–98.
8. Radenberg T, Scholz P, Bergmann G, Mayr GW. The quantitative spectrum of inositol phosphate metabolites in avian erythrocytes analysed by proton n.m.r. and h.p.l.c. with direct isomer detection. *Biochem J* 1989;264:323–333.

9. Koppitz B, Vogel F, Mayr GW. Mammalian aldolases are isomer selective high-affinity inositol polyphosphate binders. *Eur J Biochem* 1986;161:421–433.
10. Phillippy BQ, White KD, Johnston MR, Tao S-H, Fox MRS. Preparation of inositol phosphates from sodium phytate by enzymatic and nonenzymatic hydrolysis. *Anal Biochem* 1987;162:115–121.
11. Campbell ID, Dobson CM. The application of high resolution nuclear magnetic resonance to biological systems. In: Glick D, ed. *Methods of biochemical analysis*. New York: Wiley, 1978;1–133.
12. Bartlett GR. Isolation and assay of red-cell inositol polyphosphates. *Anal Biochem* 1982;124:425–431.
13. Mayr GW. A novel metal dye detection system permits picomolar-range h.p.l.c. analysis of inositol polyphosphates from non-radioactively labelled cell or tissue specimens. *Biochem J* 1988;254:585–591.
14. Lanzetta PA, Alvarez LJ, Reinach PS, Candia OA. An improved assay for nanomole amounts of inorganic phosphate. *Anal Biochem* 1979;100:95–97.
15. Jeener J. Unpublished presentation at Ampere International Summer School II, Basko Polje, Yugoslavia 1971.
16. Mayr GW. *Inositol phosphates: structural components, regulators and signal transducers of the cell – a review*. Topics in Biochemistry. Mannheim: Boehringer, 1988.
17. Dabrowski J. Application of two-dimensional NMR in the structural analysis of oligosaccharides and other complex carbohydrates. In : Croasmun WR, Carlson RMK, eds. *Two-dimensional NMR spectroscopy*. New York: VCH Publishers, 1987;349–386.

Methods in Inositide Research,
edited by Robin F. Irvine.
Raven Press, Ltd., New York © 1990

7

Mass Determination of Inositol Phosphates by High-Performance Liquid Chromatography with Postcolumn Complexometry (Metal–Dye Detection)

Georg W. Mayr

Institute for Physiological Chemistry, Faculty of Medicine, Ruhr-University Bochum, D-4630 Bochum 1, FRG

Most data published on agonist-induced cellular inositol phosphate responses have been obtained as radioactivity changes after prelabelling of cells or tissue specimens with *myo*-[^{3}H]inositol or [^{32}P]phosphate. Due to the inability to reach equilibrium labelling of all inositol phosphates, these radioactivity data in most cases preclude estimating concentrations and changes therein in terms of masses.

Therefore, methods have been developed over the past few years that allow a highly sensitive direct determination of the masses of inositol phosphates. For such analysis, there are two basic prerequisites. (i) Due to the micromolar physiological concentrations of inositol phosphates, the amount detectable should be in the picomolar range or below. (ii) As the number of inositol phosphate isomers identified in any cell or tissue is steadily increasing, such an analysis must be highly isomer or even enantiomer specific.

Two basic techniques are employed. The first is to determine directly any particular inositol phosphate isomer in an extract without preceding chromatographic separation from other isomers, e.g., by a radioligand competition assay employing highly specific high-affinity binding proteins or by enzymatic assays. The advantage of such a technique is that it does not require much instrumentation. However, up to now only one such assay has been developed to sufficient reliability, namely that for *myo*-inositol 1,4,5-trisphosphate, which employs a binding protein from adrenal cortex (1) (see Chapter 10). In future, the availability of new binding proteins (2) or antibodies highly specific for other inositol phosphates might increase the number of isomer-specific radioligand binding assays. Likewise, the availability of highly specific inositol phosphate kinases might facilitate enzymatic assays without preceding isomer separation.

In the second technique, inositol phosphate isomers are first separated chromatographically, most commonly by anion exchange high-performance liquid chromatography (HPLC), and subsequently their masses are determined by different methods: for example, off-line dephosphorylation of inositol phosphates with subsequent analysis of released P_i by microspectrophotometry (3) or of released *myo*-inositol by gas chromatography (4) or microfluorometry (5). These techniques, two of which are described in this volume (see Chapters 8 and 9), are suitably sensitive but very laborious.

On-line detection or postcolumn derivatization methods have therefore been tried as alternatives. The classical technique for detection of phosphoesters, acid hydrolysis in a long heating coil with subsequent phosphomolybdate analysis, is complicated by the stability of most inositol phosphates under acid conditions. Remaining incompletely dephosphorylated inositol polyphosphates cause strong interference with the phosphomolybdic complex (my own observations and see (6)). Even a significant improvement by Meek (7), who employed a phosphatase-loaded postcolumn reactor, could not eliminate this problem: peak heights for higher inositol phosphates were still severely depressed or even negative. Furthermore, the dead volume of the postcolumn reactor led to a critical peak broadening, a loss of activity of the postcolumn reactor posed calibration problems, and the malachite green technique (see Chapter 1) for P_i analysis was of limited sensitivity.

Amperometric detection and indirect absorbance or refractive index detection [see (8) for a broader discussion of these techniques] in principle are adoptable for on-line detection of inositol phosphates. They might be successfully employed when inositol phosphates are isocratically separated by anion exchange or ion-pairing chromatography [see (8) and Chapter 5] but pose severe sensitivity problems with the standard elution systems for inositol phosphates – salt gradients on strong anion exchangers. Electronic suppression is insufficient in this case, but the recently developed chemical suppression techniques employing suppressor columns (9,10) or a micromembrane suppressor (11) are more promising (see Chapter 11). However, the strongly basic eluants which are required for these techniques cause severe peak broadening, column memory phenomena, and reduced recoveries in the case of higher inositol phosphates (my own unpublished data). The very sensitive method of pulsed amperometric detection (PAD) (12) is restricted to carbohydrates containing two vicinal hydroxyl groups, a condition not fulfilled by many inositol phosphates. Due to this restriction and again to the strongly alkaline eluants necessary, PAD is of only limited value here.

Gas chromatography–mass spectrometry (GC–MS) has been tried extensively in Sherman's group for mass analysis of inositol phosphates. Although a sufficiently sensitive and isomer selective analysis seems to be feasible for inositol mono- and bisphosphates, derivatization and detection problems still preclude a satisfactory determination of inositols with higher phosphate content (13). Recently, GC with flame ionization detection (FID) was also successfully employed for trimethylsilylated inositol monophosphates (14).

An intriguingly simple and versatile on-line detection technique not requiring any chemical derivatization of the HPLC-separated inositol phosphates and working with a standard ultraviolet/visible wavelength (UV/VIS) HPLC-detector has recently been developed in my laboratory. The technique, termed metal–dye detection (MDD), is applicable for all inositol phosphates of physiological interest as well as for deacylation products of phosphoinositides. It is a kind of "postcolumn complexometry" in which a transition metal and a metal-specific dye act as reporter substances for inositol phosphates and other (oxy)anions eluting from the HPLC column.

Specificity for sugar and inositol phosphates on this postcolumn complexometry is achieved by choosing a suitable "detector metal" (e.g., yttrium) that forms stable complexes with phosphate groups, by removing nucleotides and, eventually, polyphosphates from the samples, and by employing a strongly acidic elution system. Thus carbonic acids are no longer retained on anion exchangers and remaining nucleotides become protonated and elute ahead of inositols of same phosphate content. A further advantage of such a strongly acidic elution system is excellent isomer selectivity for inositols containing three or more phosphate groups. Further aspects and validations of the technique have been published (15,16,17).

CHEMISTRY OF METAL–DYE DETECTION

Metal–dye detection is a postcolumn derivatization method that does not involve any covalent modification of the substances to be detected; it is simply a kind of "on-line complexometry". The ternary complexometric detection reaction, continuously brought about in the column effluent, involves the anions to be detected and two reporter substances. The latter ideally should be: (i) a chemically stable metal ion, termed detector cation, which rapidly (within a few seconds) forms very high-affinity complexes (K_d values below the nanomolar range) only with phosphorylated substances, and (ii) a reporter dye "sensing" the degree of complexation between the detector metal and the anions of interest. Ideally, the latter compound is chemically stable, has an affinity for the detector cation that is about two orders of magnitude lower than those of the anions to be detected, binds only the detector cation in a rapidly reversible pH-independent equilibrium, and produces a large and specific optical signal as a result of this complexation. Since protonation generally weakens metal complexes with phosphoester-type (oxy)anions, the whole reaction must take place at pH values near or above neutrality. Figure 1 depicts the two competitive metal binding

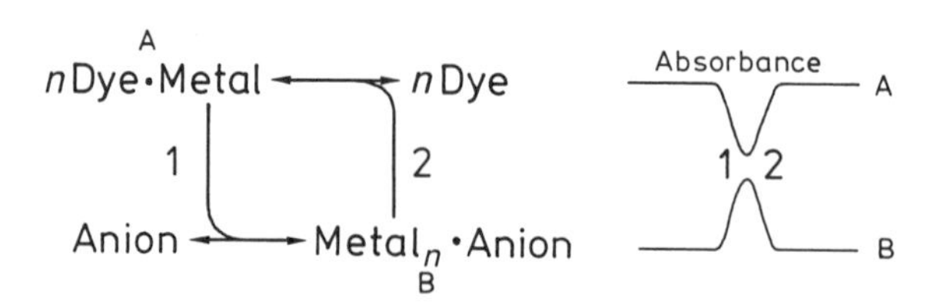

FIG. 1. The principle of metal–dye detection. The reactions depicted take into account a 1:1 stoichiometry of the dye and the detector metal in the complex, as is the case for PAR and yttrium. On the right-hand side the changes in absorbance, i.e. in the amount of metal–dye complex A, are depicted (upper trace). These occur when a peak of metal-complexing anion appears in the column effluent. Note that the basal absorbance level is high and the negative absorbance peak observed follows precisely the formation of metal–anion complex B (lower trace). (Reproduced with permission from ref. 26.)

reactions taking place and their potential effects on absorbance. A detection system meeting all above-mentioned criteria has not been found yet. Problems encountered in the process have been: the redox-stability of detector metals; the specificity of a metal for phosphocompounds; the tendency of metal–oxyanion complexes, metal hydroxides and metal–dye complexes to precipitate above neutrality; the chemical instability and light sensitivity of reporter dyes; the specificity of dyes for the detector metal; the insufficient purity of dyes and their weak affinity for the metal; and interference from solvent or buffer contaminants. The last of these complications can result both from contaminating metals forming complexes with the phosphocompounds or with the dye and from contaminating anions binding detector metal. In any of these cases there may be loss of detection sensitivity and, sometimes, severe baseline drifts and deviations from detection linearity. None the less, a careful circumvention of potentially interfering conditions has led us to several successes with on-line detection for inositol phosphates.

The system which we have been employing continuously now for more than two years on two automated setups is described here in detail. It employs yttrium as the detector cation and 4-(2-pyridylazo)resorcinol (PAR) as the reporter dye, leading to detection of picomole amounts of phosphocompounds by using negative absorbance peaks at 520 to 550 nm. It should be pointed out, however, that detection can be achieved in many more ways and for many more purposes, e.g., by using a fluorescent dye or a detector cation suitable for detecting specifically other types of anions, for example sulphocompounds. The system described here, though not the most sensitive one found and not fulfilling all the criteria mentioned above, has a number of practical advantages. Yttrium, a trivalent transition metal, binds to phosphorylated compounds with high affinity when the pH is above 7. K_d values for most inositol phosphates are in the nanomolar range. Unlike many other metals it is redox-stable and thus does not destroy the column matrix. Its tendency to form insoluble hydroxycomplexes is minor below pH 9. Complex formation with sulphocompounds is weaker than with phosphocompounds. Among all lanthanides and transition metals, yttrium forms a complex with PAR which has almost the highest stability constant. The use of PAR as a reporter dye has the following further advantages, although its metal-specific absorbance is dependent on pH: the pure material is available at a low price; it exhibits a high selectivity for polyvalent cations, like yttrium, with ionic radii of about 0.9 Å (1 Å = 0.1 nm) (16); the fully metal-specific absorbance peak at 520 nm is very high (with yttrium up to 60,000 M^{-1}, depending on pH (18)); monovalent alkali cations are only weakly bound and large monovalent organic cations, as employed in the eluants and the derivatization reagent (see below), are not bound at all; when protected from light, metal-free aqueous solutions between pH 5 and 9 are stable for several weeks at room temperature.

The eluants most frequently employed for separation of inositol phosphates by anion exchange HPLC, ammonium formate or phosphate adjusted to pH 3.7 with phosphoric acid, are of course incompatible with MDD. But all eluants devoid of phosphate, sulphate and bicarbonate can be employed. Eluants containing formate, acetate, trifluoroacetate, chloride or perchlorate have been successfully employed in my laboratory. Optimal counterions are monovalent large organic cations (e.g., triethylammonium, triethanolammonium, trishydroxymethylammonium, tetramethylammonium or imidazolium), but ammonium, potassium and low concentrations of sodium or lithium can be employed also. Only strongly acidic eluants can simultaneously contain the detector cation, since at higher pH values complex formation with inositol phosphates interferes severely with separation of the latter. When the detector cation is not included in the eluants but in the postcolumn derivatization reagent, the pH of the latter must also be slightly acidic (about 5), since otherwise insoluble hydroxycomplexes and mixed complexes with the dye are formed slowly.

In practice, two appropriate variants of MDD can be achieved with a two-pump HPLC system with low-pressure gradient formation. In variant I, the detector cation is included in a strongly acidic eluant and the derivatization solution contains the dye and sufficient buffer capacity to bring the pH of the effluent to above neutrality. This variant is most frequently employed in our laboratory. In variant II, the pH of the eluant is around neutrality. Both the detector cation and the dye are contained in a slightly acidic postcolumn derivatization reagent. The buffer capacity of the eluant is sufficient to maintain a final pH around or above neutrality.

CHEMICALS, STANDARDS, AND COLUMNS

Chemicals

Analytical grade HCl (lots are individually tested for purity) is from Riedel de Haen (Seelze, FRG), perchloric acid from Baker. Triethanolamine of

>99% purity is from Fluka (Buchs, Switzerland) or Serva (Heidelberg, FRG). PAR (p.a.) and Norit A (p.a.) are also from Serva. PAR from Sigma (St Louis, MO) is not pure enough. Yttrium trichloride hexahydrate is from Aldrich (Milwaukee, WI). Tetramethylammonium hydroxide and tetramethylammonium chloride are from Janssen (Beerse, Belgium). A 20 mM stock solution of PAR is prepared in methanol (p.a.) and stored at −30°C. Norit A is further treated by boiling for 2 h in 3 M HCl, washing with water to neutrality and drying at 120°C.

Standards

Inositol monophosphates (Ins(1)*P*, Ins(4)*P*), bisphosphates (Ins(1,4)P_2, Ins(2,4)P_2, Ins(4,5)P_2), trisphosphates (Ins(1,4,5)P_3, Ins(2,4,5)P_3), tetrakisphosphate (Ins(3,4,5,6)P_4), and pentakisphosphate (Ins(1,3,4,5,6)P_5), glycerophophoinositol monophosphate (Gro*P*Ins(4)*P*) and bisphosphate (Gro*P*Ins(4,5)P_2), fructose bisphosphate (Fru(1,6)P_2), 3-phosphoglycerate, α-glycerophosphate, and phosphoenolpyruvate (PEP) are from Boehringer Mannheim (FRG). Sodium phytate, glucose bisphosphate (Glc(1,6)P_2), 6-phosphogluconate, 2,3-bisphosphoglycerate, sedoheptulose 1,7-bisphosphate, 1-phosphoribosyl 5-pyrophosphate, polycarbonic acids, phosphocarbonic acids, sodium pyrophosphate, trimetaphosphate and polyphosphate fractions are from Sigma. Glycerophosphoinositol (Gro*P*Ins) is prepared by deacylation of phosphatidyl inositol (PtdIns) (19); Ins(1,3,4,5)P_4 is prepared by enzymatic phosphorylation of Ins(1,4,5)P_3 (20) and subsequent purification by anion exchange chromatography (see Chapter 2). Further *myo*-inositol phosphate isomers[1] are isolated from partly hydrolysed phytic acid (15,21) or from an avian red blood cell extract (22) by anion exchange chromatography on 200 cm × 1.5 cm columns of Q-Sepharose (Pharmacia/LKB, Uppsala, Sweden). Conditions for inositol phosphate isomer separation and quantification are described in more detail below and by Radenberg et al. (16). All isolated inositol phosphates and Gro*P*Ins are checked routinely for their isomerism and purity by proton and ^{13}C nuclear magnetic resonance (see Chapter 6).

Columns

Mono Q anion exchange beads (not available in loose form) are purchased as Mono Q HR 10/10 columns from Pharmacia/LKB and refilled into empty HR 5/5 and HR 20/5 columns (Pharmacia/LKB) or 4 mm × 50 mm and 4 mm × 300 mm PEEK columns (NO-MET columns, Chromatographie

[1] Ins*P*s obtained by hydrolysis of phytic acid will be racaemic mixtures. Thus, in Fig. 5 and Table 1 Ins means D- or L-*myo*-inositol.

Handel Mueller, Fridolfing, FRG). For column packing, a 1:1 slurry (wet packed beads to water) is made and the columns filled using a disposable pipette with a long 3 mm external diameter tube attached, facilitating a bubble-free introduction of the slurry. Packing is accelerated by pumping water through the partially packed column at a flow rate of 0.2 ml/min. Finally, the flow rate is increased for some minutes until the back pressure reaches about 50×10^5 Pa. For some separations, a prepacked Propac PA1 column (4 mm × 250 mm, Dionex P/N 039658) is employed without a guard column.

INSTRUMENTATION AND SYSTEM SETUP

The basic setup of an automatized HPLC system suitable for an inositol phosphate microanalysis with MDD is depicted in Fig. 2.

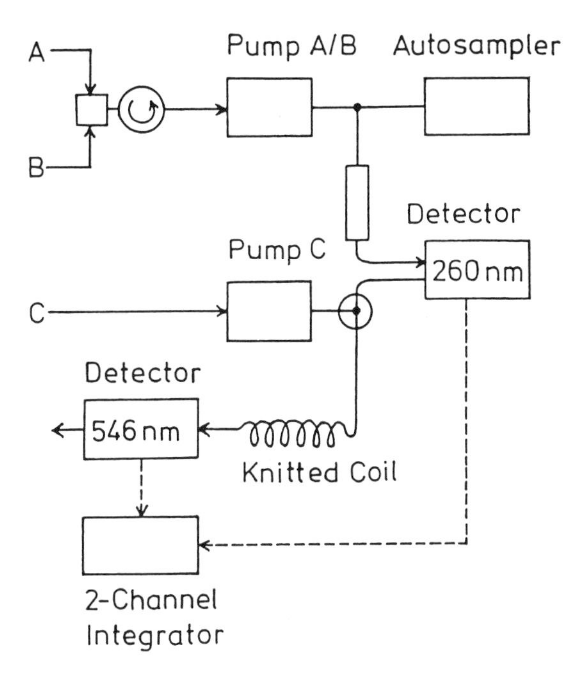

FIG. 2. Scheme of an HPLC system suitable for metal–dye detection. Note that the two absorbance signals cannot be recorded by one two-channel monitor. A, B and C denote the two variants (I and II, see Elution systems) of weak eluant, strong eluant, and postcolumn derivatization reagent, respectively. (Reproduced with permission from ref. 26.)

Gradient Formation

Low-pressure gradients for the routinely employed anion-exchange separations are formed by means of a three-way magnetic switch-valve (Fa. Lee, Frankfurt, FRG) and a magnetically stirred mixing chamber of 1–2 ml dead volume (LKB Ultrograd mixer). An additional small bubble trap is installed after the mixer. Although retention times are increased by this setup, one HPLC pump is thus free of delivering the derivatization reagent.

System Control

To control the system an LCC 500 plus or a 2152 controller is employed (both from Pharmacia/LKB). Any other controller might be appropriate if it can control the gradient mixing valve and the flow rate of the solvent delivery HPLC pump. It should also give appropriate signals to control an automatic sample injector or a motor driven injection valve, to start two integrator channels separately (see below), to set one monitor to zero, and to switch a second HPLC-pump on and off, or even better, to control its flow.

Solvent Delivery

For delivery of both the solvent and the postcolumn reagent, true double-piston pumps with (identical) stroke volumes below 80 μl are essential. With single-piston pumps and pumps having larger stroke volumes severe oscillations of the high basal absorbance resulting from the coloured derivatization solution occur due to flow pulsations. Pumps with spindle-driven pistons (e.g., fast protein liquid chromatography pumps) give rise to even worse flow oscillations. One and a half piston pumps may produce sufficiently smooth flow, but baseline oscillations due to interference between the mixing-valve switching and the solvent aspiration periods have been encountered. After testing a number of HPLC pumps, we have selected the 2150-type pumps from Pharmacia/LKB. Their piston seal rinsing system prolongs their seal lifetimes in high salt solvents. Prior to use, solvents and the postcolumn reagent are filtered and degassed by vacuum filtration through inert 0.22 μm pore size membrane filters (Millipore, type GV) in a glass filtration device. Teflon filter frits (Pharmacia/LKB) are used to cover the ends of the solvent delivery lines. The eluants, but not the derivatization reagent, are continuously degassed by a small stream of helium without pressurization. For the low-pressure gradient formation, slightly positive equal filling pressures in both eluant lines are optimal. They are brought about by gravity by adjusting both eluant reservoir levels to about 30 cm above the pump level. Since in elution variant I the derivatization reagent is relatively viscous (see composition below), a positive hydrostatic filling pressure (50 cm) is also applied to this pump.

Wetted Parts

As higher inositol phosphates tend to adsorb on to steel surfaces (phytic acid is an anticorrosive! (23)), and since up to 0.5 M HCl is employed as eluant, all wetted system parts should be inert. We use the inert versions of the 2150 HPLC pumps. All fittings, filter units, column inlets and column frits are made from inert polymers. Columns are made from glass or polymer (see above) and Teflon tubing of 0.5 mm internal diameter and 1.7 mm external diameter (resisting pressures up to 100×10^5 Pa) is employed throughout the whole system.

Sample Injection

An automatic sample injector is highly recommended for two reasons: (i) baseline instabilities are strongly reduced by precisely iterated injection–elution cycles; and (ii), more importantly, the system can be operated day and night. Any automatic sampling device in which parts to be wetted are inert or can be substituted with an inert substance is adequate. Maximal injection volumes should be 2 ml. Sample loops must also be of Teflon tubing. We employ an ACT 100 automatic injector (Pharmacia) equipped with a high-pressure motor valve PMV 7 and a 2157 Autosampler (LKB) in which all wetted parts except the injector needle have been substituted with inert ones. When an autosampler is not available, an inert motor-driven injection valve (such as the PMV 7 mentioned above) should be employed instead of manual injection.

Column Assembly

A Mono Q HR 20/5 column (see above) with a Mono Q HR 5/5 column as guard column is routinely employed for physiological samples. An inert self-constructed in-line filter containing a sandwich of two 8 mm diameter Whatman no. 1 cellulose discs with an inert 1 μm filter disc between (type TE 37, ref. no. 411099, cut by Schleicher and Schuell) protects the column from charcoal and other fines. Similar inert in-line filter units are commercially available from DIONEX and Chromatographie Handel Mueller (above). When shorter columns (25–100 mm length) are operated for rapid analyses, only the in-line filter is employed for column protection.

Postcolumn Derivatization

For mixing the postcolumn derivatization reagent with the column effluent, a T-shaped three-way junction of low dead-volume, and a short knitted coil (1 mm internal diameter Teflon tubing) with a dead volume of 400–500 μl are employed. Fines in the postcolumn reagent are removed by another in-line filter unit (Pharmacia/LKB) loaded with an 8 mm diameter Whatman no. 1 cellulose disc which is installed before the mixing T.

Monitors

A system employed for physiological samples should be equipped with two absorbance monitors in series. The first (optional) detector should be a standard UV detector with an inert semimicro- or micro-flow-through cell (dead volume < 20 μl) set to 260 nm and connected directly to the column outlet prior to postcolumn derivatization. It allows for detection and discrimination of peaks caused by remaining nucleotides (which are also detected by MDD, see above). The second detector is a VIS detector with a flow-through cell of similar dead volume, which is set to 520 or 546 nm. The MDD signal characterized by a high basal absorbance is monitored by this detector. In order to obtain highly reproducible starting baseline levels, the detector must be equipped with an externally controlled auto-zeroing device. Due to the greater noise implicit in the recorded signal, this monitor should allow for a signal integration with time constants between 1 and 4 s.

Signal Integration

The absorbance signals from both detectors – that from the VIS detector after inversion (interchange of the plus and zero/minus lines) – are integrated by a two-channel integrator which should possess the option for subtracting a baseline and optimally should allow for an averaging of several baselines or for their numerical smoothing. Both a Shimadzu CR5A integrator and a Nelson chromatography system are used in our laboratory. The inverted VIS detector signal should be strongly attenuated in the monitor or, if this is not possible, by a voltage dividing unit, since a negative "system peak" early in the gradient may pose problems due to the input offset of some integrators (e.g., Spectraphysics integrators). By starting the UV-recording channel before the VIS channel the two recordings can be matched exactly.

Optional Column Switching Setup

For a very rapid analysis of only one or a limited number of inositol phosphate isomers, e.g., for determination of inositol phosphate kinase or phos-

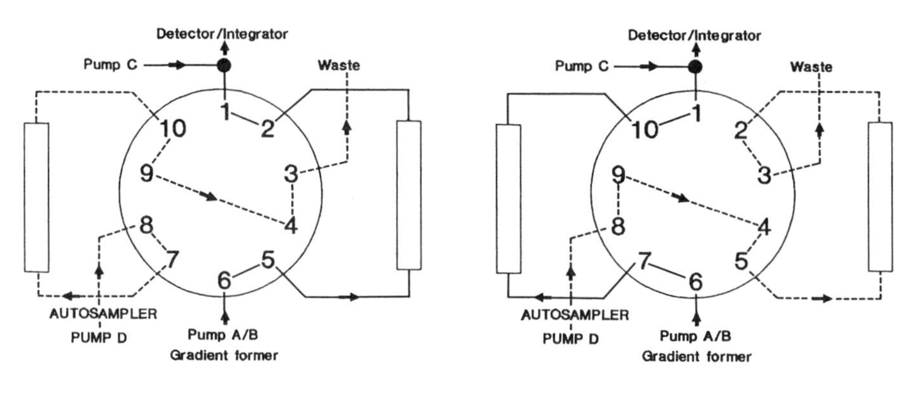

FIG. 3. Column switching setup for alternating operation of two columns in one HPLC system. The kind of connection of the 10-way switch valve to the system components and the two identical separation columns is depicted. At both valve positions one of the two columns is regenerated or loaded, respectively, through the action of a third pump (pump D), while the other is subjected to gradient elution. Corresponding connections are indicated by broken and continuous lines, respectively.

phatase substrates and products, target bound inositol phosphates, or deacylated phosphoinositides, rapid elution protocols can be employed on short Mono Q columns (e.g., 5 mm × 25 mm). In order to increase further the frequency of these analyses we employ a column switching setup, by which two identical columns are operated alternately. While one column is submitted to gradient elution with MDD monitoring, the second one is regenerated and subsequently sample-loaded and vice versa. When the elution does not take much longer than the latter two phases, the time required for one cycle is halved. Only two inexpensive system components are additionally required: an externally controlled motor-driven or pneumatic 10-way valve (or two simultaneously switched 6-way valves; we are employing a motor-driven 10-way valve from VALCO Instruments) and a simple one-piston HPLC pump. The latter should be equipped with an electrical on/off switching device. Since these components are only wetted by the low salt eluant they need not be inert. Figure 3 shows how the 10-way valve is connected to the columns and the system components; Figure 4 graphically depicts the essential controller actions and the short gradient which we employ for separating Ins(1,4,5)P_3 and Ins(1,3,4,5)P_4 obtained from non-radioactive Ins(1,4,5)P_3 3-kinase assays.

Optional On-Line Radioactivity Detection

The mixture of effluent with postcolumn derivatization reagent can be passed directly through an appropriate flow-through radioactivity detector (we em-

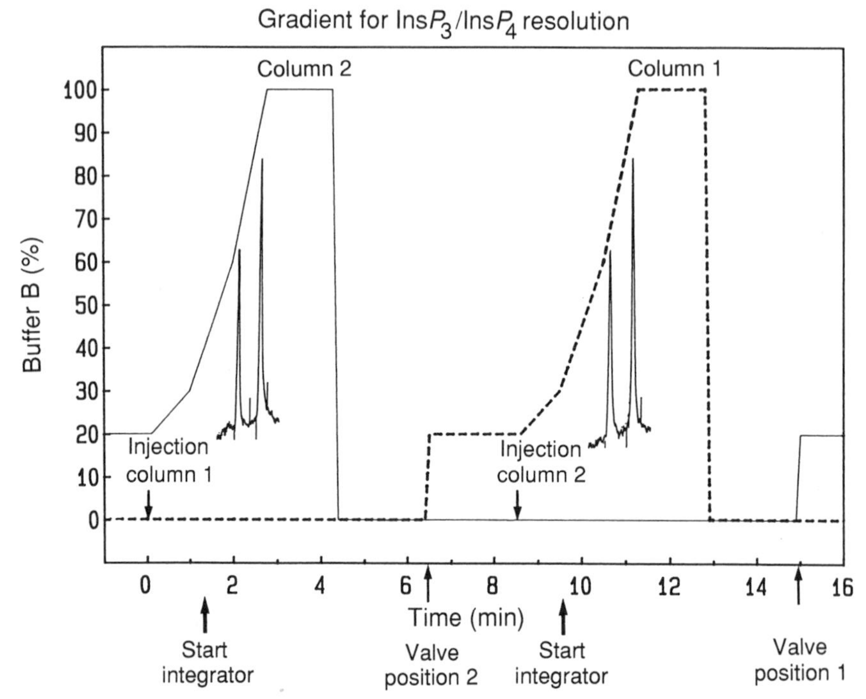

FIG. 4. Working protocol for an HPLC system with two-column switching setup. The column switching assembly shown in Fig. 3 is employed here for separation of Ins(1,4,5)P_3 and Ins(1,3,4,5)P_4, the substrate and product of a non-radioactive Ins(1,4,5)P_3 3-kinase assay. Two complete analyses – each requiring 8 min – are performed within the duration of one column switching cycle. Within this cycle, two gradients are applied (each on a different column), and two chromatograms are recorded. Eluant A is as described for elution system I, eluant B is 0.2 M HCl, containing 15 μM YCl_3. The flow rate of pump A/B (eluant delivery) and the additional pump D (column reconditioning/sample loading) are 3 ml/min, the postcolumn reagent (reagent C of elution system I) is pumped at 1.5 ml/min. Two calibration runs with 1 nmol each of Ins(1,4,5)P_3 and Ins(1,3,4,5)P_4 injected (in 1 ml) are shown on the inserted chromatograms.

ploy an HPLC radioactivity monitor LB 506 C1 from Berthold). In the slightly coloured effluent, the [^{32}P]phosphate radioactivity can be monitored directly by Čerenkov radiation and the [^{3}H]inositol radioactivity monitored after continuously adding an appropriate volume of a liquid scintillator suitable for salt-containing aqueous solutions (e.g., Zinsser no. 306 scintillator at a volume ratio 3:1).

ELUTION SYSTEMS

Variant I (Strongly Acidic Elution System)

Weak eluant (solvent A): 0.2 mM HCl, 15 μM YCl_3
Strong eluant (solvent B): 0.5 M HCl, 15 μM YCl_3

Postcolumn reagent (reagent C): 1.6 M triethanolamine·HCl (pH 9.0), 300 μM PAR

Comments

Weakly acid-labile compounds such as nucleoside polyphosphates, polyphosphates, most hexose bisphosphates, and glycerophosphoinositol phosphates are not destroyed during the elution with this eluant. More acid-labile compounds such as inositol cyclic phosphates, phosphoribosyl pyrophosphate, fructose 2,6-bisphosphate, and acylphosphates are partly destroyed during elution with this eluant. Although the columns are briefly exposed to a pH value of about 0.4 at the end of a gradient, the styrol-based column materials employed (see above) are sufficiently stable to maintain their performance for more than 4000 runs (the maximal number employed up to now). Silica-based anion exchangers are not sufficiently acid resistant!

Variant II (Neutral Elution System)

Weak eluant (solvent A): 10 mM triethanolamine·HCl (pH 7.5)
Strong eluant (solvent B): 10 mM triethanolamine·HCl, 1 M tetramethylammonium chloride (pH 7.5)
Postcolumn reagent (reagent C): 300 μM PAR, 45 μM YCl_3 2mM ammonium acetate, pH 5.0.

Comments

The eluant pH may be varied between 6.5 and 8.5. Variant II leads to significantly higher detection sensitivities for higher inositol phosphates than does variant I. It is highly recommended for acid-labile inositol phosphate derivatives such as phosphoinositide deacylation products (see example below), cyclic inositol phosphates or chemically synthesized acid-labile derivatives. Its disadvantage is that nucleotides and inositol phosphates with the same phosphate content partially interfere because of their elution positions. Furthermore, remaining polycarboxylic acids such as oxalate, citrate, EDTA or EGTA, are bound under these conditions and can also interfere with inositol phosphate peaks. When the concentration of tetramethylammonium in solvent B is 1 M, inositol compounds containing up to four phosphate groups are eluted; when it is 1.6 M, compounds containing up to seven phosphate groups are eluted. In comparison with variant I, the isomer selectivity of this system is better for inositol phosphates containing one and two phosphate groups but worse for inositol phosphates containing four to six phosphate groups. The system is suitable as an alternative for isomer assignments by co-chromatography with defined standards.

CONTROLLER PROTOCOLS

Protocol I (Long Column Gradient for Elution Variant I)

The gradient for separation of InsP_2 to InsP_6 isomers from physiological extracts on a 5 or 4 mm diameter column, respectively, of total length (including 50 mm guard column) of 250 or 350 mm (see above) by elution variant I is:

0 min: 0% B; flow A/B 1 ml/min; flow C 0 ml/min; autosampler ready; loop (n) times;
0.2 min: flow A/B 1.6 ml/min;
4 min: start sample loading (hold gradient for 0.5 to 2 min, depending on sample loop volume which is filled at 1 ml/min); start injection (back to load position after 5 min);
7 min: flow C 5 ml/min;
7.2 min: flow C 0.8 ml/min;
9 min: flow A/B 1 ml/min; flow C 0.5 ml/min;
11.7 min: start integration of UV detector signal;
11.8 min: auto zero of VIS monitor;
12 min: start integration of VIS detector signal; 4% B;
15 min: auto-zero of VIS monitor (optional);
22 min: 8% B;
31 min: 14.5% B;
44 min: 28% B;
52 min: 44% B;
56 min: 60% B;
61 min: 90% B:
64 min: 100% B;
71 min: 100% B;
71 min: 0% B; (reset autosampler);
74 min: flow C 0 ml/min; flow A/B 1.6 ml/min;
75 min: end of loop (next cycle).
After last cycle: continue flow A/B at 1.6 ml/min; 100 min: stop.

An application of this protocol is shown in Fig. 5.

Comments

For shorter columns, the column preconditioning/regeneration, washing and gradient elution phases can be appropriately shortened. Furthermore, flow rates during column preconditioning/regeneration can be increased up to 3 ml/min and 2 ml/min for column lengths of 50 and 100 mm, respectively. The flow rates of the derivatization reagent should be appropriately adopted. Upward concave gradients are optimal for a resolution of all inositol phos-

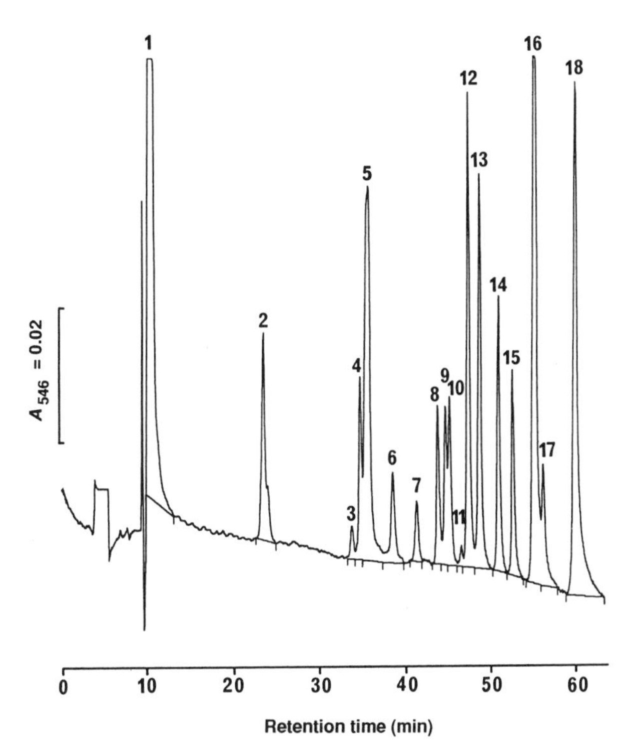

FIG. 5. HPLC–MDD analysis of inositol phosphates — elution protocol I. A Mono Q HR 20/5 column with guard column Mono Q 5/5 was employed with elution system I and protocol I (see the text). A partial hydrolysate of phytic acid containing between 39 (Ins(1,3,4,5)P_4) and 1970 (Ins(1,2,4,5,6)P_5) pmol of individual inositol phosphate isomers was injected. Resolved peaks even when containing more than one isomer are labelled by one isomer name only: 1, Ins$P$ + P_i; 2, Ins(1,4)P_2; 3, Ins(1,3,5)P_3; 4, Ins(1,3,4)P_3; 5, Ins(1,4,5)P_3; 6, Ins(1,5,6)P_3; 7, Ins(4,5,6)P_3; 8, Ins(1,2,3,5)P_4; 9, Ins(1,2,3,4)P_4; 10, Ins(1,2,4,5)P_4; 11, Ins(1,3,4,5)P_4; 12, Ins(1,2,5,6)P_4; 13, Ins(2,4,5,6)P_4; 14, Ins(1,4,5,6)P_4; 15, Ins(1,2,3,4,5)P_5; 16, Ins(1,2,4,5,6)P_5; 17, Ins(1,3,4,5,6)P_5; 18, InsP_6. See Table 1 for more details.

phates. For a stable long-term performance of the postcolumn derivatization system it is important to stop the flow of reagent C periodically. Optimally, this is done while strong eluant is still leaving the column and maintained until shortly before starting the gradient again (see protocol above). Without this manoeuvre a colour-coating of the knitted coil may finally extend into the VIS detector cuvette and pose severe baseline problems. This phenomenon is a sensitive indicator of impurities in the PAR. When samples are loaded manually, the durations of cycle phases should be kept constant.

Protocol II (Long Column Gradient for Elution Variant II)

The gradient for separation of Ins$P$ to InsP_4 isomers and of Gro$P$Ins to Gro$P$InsP_3 isomers from physiological extracts on a 5 or 4 mm diameter columne, respectively, of total length (including 50 mm guard column) of 250 or 350 mm (see above) by elution variant II is:

0 min: 0% B; flow A/B 1 ml/min; flow C 0 ml/min; autosampler ready; loop (n) times;
6 min: start sample loading (hold gradient for 0.5 to 2 min, depending on sample loop volume, which is filled at 1 ml/min); start injection (back to load position after 4 min);
7 min: flow C 5 ml/min;
7.2 min: flow C 0.5 ml/min;
9.7 min: start integration of UV detector signal;
9.8 min: auto-zero of VIS detector;
10 min: start integration of VIS detector signal;
10 min: 4% B;
15 min: 8% B;
21 min: 14.5% B;
30 min: 28% B;
35 min: 44% B;
38 min: 60% B;
42 min: 90% B;
44 min: 100% B;
50 min: 100% B;
50 min: 0% B: (reset autosampler);
54 min: flow C 0 ml/min; end of loop (next cycle).
After last cycle: continue flow A/B at 1.0 ml/min; 80 min: stop.

An application of this protocol is demonstrated in Fig. 6.

Comments

This protocol is optimally suited for separation of Gro*P*InsP_x isomers (see Fig. 6) obtained from deacylated phosphoinositides. When 1.6 M tetramethylammonium is employed for a separation of all inositol phosphate isomers, a longer gradient should be used. The time required for column reconditioning is shorter with elution variant II than with variant I, since only in the latter case has the proton buffering capacity of the column to be overcome. For shorter columns, shorter elution protocols can be employed (see above), but again the optimal gradient profiles are upward concave. With longer isocratic phases of eluant A containing low concentration of tetramethylammonium chloride (20–100 mM) or with shallow linear gradients, mono- and bisphosphorylated inositol phosphates can be resolved with high isomer-selectivity.

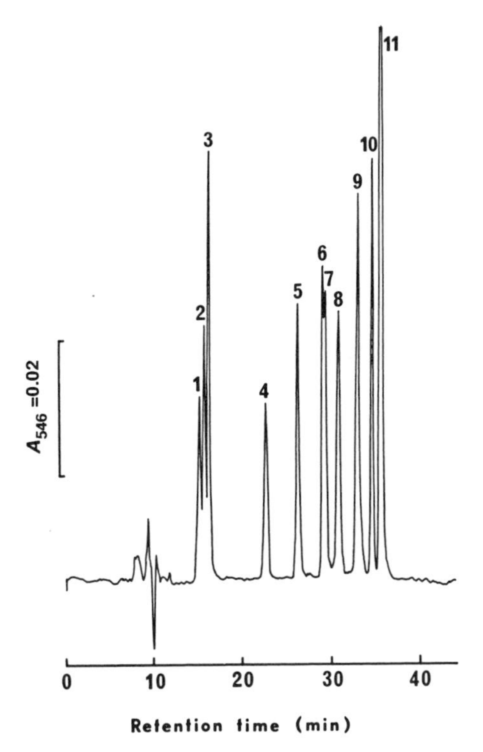

FIG. 6. HPLC–MDD analysis of deacylated phosphoinositides and other compounds – elution protocol II. A Mono Q HR 20/5 column with guard column Mono Q 5/5 was employed with elution system II and protocol II (see the text). Peaks numbered are (quantities injected in nmol): 1, Ins(4)*P*(10); 2, αGro*P*(10); 3, P_i (20); 4, Gro*P*Ins(4)*P*(10); 5, 3-phosphoglycerate (2); 6, Glc(1,6)P_2 (2); 7, Fru(1,6)P_2 (2); 8, Ins(1,4)P_2 (2.5); 9, Gro*P*Ins(4,5)P_2 (3); 10, 2,3-bisphosphoglycerate (2); 11, Ins(1,4,5)P_3 (3).

SYSTEM MAINTENANCE

A number of parts should be exchanged in regular intervals in order to guarantee an optimal system performance.

Knitted coil: every week of continuous operation; colour-coating, if occurring after some days, can be removed by purging with 10–20 ml of 2 M HCl in ethanol and then water (through the T-shaped junction after disconnecting it from the column).

In-line filters: every week the filter in the reagent delivery line, every second week the filter sandwich protecting the column.

Column inlet frits: every two months.

Teflon tubing: lines wetted only by weak eluant, water, or sample solution every four months.

Solvent aspiration frits: for postcolumn reagent every month, for eluants every six months.

Reservoir and aspiration line for reagent C: cleaning every month by rinsing with 0.1 M HCl.

Columns: when the system is stopped the columns can be washed and kept in the weak eluant of variant I. A few millimetres of filling at the top of the inlet column should be carefully removed every four months and fresh beads substituted.

STANDARDIZATION AND CALIBRATION

Quantification of Standards

Unfortunately, there is no set of precisely quantified inositol phosphate standards available. Therefore, precise quantifications of all inositol phosphates employed as standards must be done by determining the organic phosphate concentration of stock solutions of these inositol phosphates. Total and inorganic phosphorus is determined as described in Chapter 6 (see also 16, 19, 21, 24). Furthermore, the isomeric purity has to be checked by HPLC with MDD. We prepared a fully quantified mixture of reference isomers by separating every resolvable inositol phosphate isomer from hydrolysates of phytic acid (containing about 3 mmol of total phosphorus) on a 1.5 cm × 200 cm column of Q-Sepharose in the chloride form (from Pharmacia/LKB). The isomer selectivity of this column is comparable to a 5 mm × 250 mm column of Mono Q. As in the case of the latter column, an upward concave gradient of 0 to 0.5 M HCl (3 litres) is employed. Every resolved peak is precisely quantified by organic phosphorus determination (see above), and the isomers present are assigned by nuclear magnetic resonance techniques (see Chapter 6). Standard solutions of these isomers and of those which are not obtained in sufficient amounts by this procedure are mixed and kept frozen in small portions. Individual component concentrations should be in the range 1–10 nmol/ml in order to facilitate the injection of precise amounts in the range 10–5000 pmol. Since some of the commercially available inositol phosphate isomers are enzymatically prepared and are not absolutely free of contaminating phosphatase activity, a standard solution, when briefly thawed, should always be kept on ice. Since within a group of inositol phosphates of defined phosphate content the detection sensitivity does not vary by more than ±10% (my unpublished data), it is not strictly necessary to have all chromatographically resolvable isomers of this group (see Table 1 for details) in the standard mixture. If a maximal absolute error of about 20% is acceptable, the peak area versus mass relationship of a neighbouring peak (corresponding to an inositol phosphate with the same phosphate content) can be employed.

TABLE 1. *Order of elution of inositol phosphates and other phosphocompounds by decreasing pH*

Ins(2)P ≤ Ins(1)P ≤ Ins(4)P ≤ Ins(5)P, **X-P**

P_i

Gluconate(6)P, glycerate(3)P

Ins(1,3)P_2 ≤ Ins(2,4)P_2 = Ins(1,4)P_2
Ins(2,5)P_2 ≤ Ins(1,5)P_2, **Glc(1,6)P_2**
Ins(1,2)P_2 ≈ Ins(4,6)P_2
Ins(1,6)P_2 ≤ Ins(4,5)P_2

Fru(1,6)P_2, sedoheptulose(1,7)P_2

PP_i, X-PP

Glycerate(2,3)P_2

Ins(1,3,5)P_3 = Ins(2,4,6)P_3
Ins(1,2,3)P_3 = Ins(1,2,4)P_3 ≤ Ins(1,3,4)P_3 ≤ Ins(1,2,5)P_3
Ins(2,4,5)P_3 ≤ Ins(1,4,5)P_3 ≤ Ins(1,2,6)P_3
Ins(1,4,6)P_3

X-PPP, PPP_i, P-Rib-PP

Ins(1,5,6)P_3

Ins(4,5,6)P_3

Ins(1,2,3,5)P_4 ≤ Ins(1,2,4,6)P_4
Ins(1,2,3,4)P_4
Ins(1,3,4,6)P_4
Ins(1,2,4,5)P_4
Ins(1,3,4,5)P_4
Ins(1,2,5,6)P_4

Ins(2,4,5,6)P_4

Ins(1,4,5,6)P_4
Ins(1,2,3,4,6)P_5

Ins(1,2,3,4,5)P_5

Ins(1,2,4,5,6)P_5
Ins(1,3,4,5,6)P_5

InsP_6

Single spacings between (groups of) peaks indicates that, on shorter columns, crowding phenomena may occur. The elution order of all 39 separable *myo*-inositol phosphate isomers is given. *X* denoted a non-hydrophobic, non-aromatic neutral residue. Non-inositol phosphates are indicated in bold.

Isomer Assignment

For assignment of each isomer, a mixture of all resolvable inositol phosphate isomers should be available. This is easily obtained from partial chemical and enzymatic hydrolysates of phytic acid, prepared as described by

Phillippy et al. (21) and supplemented with Ins(1,3,4)P_3, Ins(1,4,5)P_3, Ins(1,3,4,5)P_4 and Ins(1,3,4,5,6)P_5, which are present in insufficient amounts in such hydrolysates. When a strongly acidic elution system is employed on different styrol-based strong anion exchange columns of low capacity, the order of elution of all *myo*-inositol phosphate isomers does not vary. This order of elution appears to be determined exclusively by the protonation properties of the individual isomers, i.e., by their individual molecular p*K* values and not by the properties of the exchange matrix. Table 1 comprises these data, which might be useful for assigning isomers by their chromatographic behaviour.

A separate set of standards should be made from other phosphocompounds frequently encountered in cell extracts, e.g., hexose monophosphates, 3-phosphoglycerate, (phosphogluconic acid), PP_i, glucose 1,6-bisphosphate, fructose 1,6-bisphosphate, (other sugar bisphosphates), and 2,3-bisphosphoglycerate. Because of incomplete chromatographic resolution, the compounds in parentheses should be kept separate. Furthermore, a mixture of the most abundant nucleotides and of inorganic tris- to pentapolyphosphates should be prepared in order to identify these compounds which potentially may cause interference. These mixtures should also be kept frozen in small portions. For an analysis of physiological samples by use of the neutral elution system, a mixture of frequently encountered (poly)carbonic acids should be prepared.

Calibration and Quantification

For most isomers peak height versus mass is linear between about 10 and 2000 pmol of isomer. Above these amounts, the slopes gradually decrease, particularly for phosphocompounds with lower phosphate content; for pyrophosphates and polyphosphates, a sigmoidal deviation from linearity (upward concavity) is observed at lower amounts. In order to allow for accurate determinations of mass, non-linear calibration routines may be employed. Since none of the commercially available integration software provides optimal techniques for this purpose, a personal computer program reading raw data for peak areas from several types of integrator and individually optimizing the non-linear calibration of every peak has been developed (''HPLC'', G. W. Mayr, unpublished). This program, which also contains useful routines for handling the enormous amount of peak data obtained by this technique and facilitating their statistical and graphical evaluation, is available upon request to the author.

SAMPLE PREPARATION

The only samples which can be injected directly are glycerophosphoinositol (phosphates), obtained by deacylation of phospholipids, and purified samples of isolated or chemically synthesized inositol phosphates or derivatives that are free from the large concentrations of carrier salts which interfere with the MDD. All aqueous extracts from cell or tissue specimens must be processed to make them suitable for MDD, i.e., undergo removal of nucleotides, and, if necessary, of carrier salts. Since the inositol phosphates present in these samples are prone to dephosphorylation by contaminating phosphatases, which appear to be present even in dust and fingerprints, all instruments used should be sterile, solutions should be made with boiled water, filtered and stored at 4°C, and gloves should be worn.

Extraction

Extractions can be performed with perchloric acid (PCA; 0.75 to 1.0 M) or trichloroacetic acid (TCA; 10% (w/v)). Cell suspensions are directly mixed with a third of their volume of 3 (to 4) M PCA containing 0.2 M acetic acid or with a corresponding volume of 40% TCA, vigorously stirred and put on ice for 30 min maximum. Weighed frozen tissue samples obtained by quick freezing techniques (e.g., freeze clamping) or by focussed microwave irradiation with subsequent immersion in liquid nitrogen are either extracted directly with 10 or more volumes (minimally 1 ml) of PCA or TCA or – in the case of collagen or fiber-rich specimens – after powdering the specimens in a liquid-nitrogen-cooled steel ball mill (Braun-Melsungen Dismembranator). Frozen powders or specimens are transferred to a 12 ml polypropylene tube and extracted using a small Ultra-Turrax homogenizer (up to five times for 10 s). After removal of precipitated protein and other cellular material by centrifugation at 5000 *g* for 5 min in the cold, the supernatants are quantitatively transferred to a second 12 ml polypropylene tube. If desired, a second extraction can be performed. For precise determination, the volume trapped in each pellet should be determined by weighing it wet and dried.

To extracts, the following additions are made:

1. An appropriate amount of an internal standard. We add routinely about 1 nmol of Ins(1,2,3,4,5)P_5 to one of two identical samples. Other inositol phosphates may be added likewise.
2. Five μmol of NaF (50 μl of a 0.1 M solution). The addition of fluoride has been found to inhibit contaminating inositol phosphatase activity.
3. Five to 20 μmol of EDTA (50 to 200 μl of a 0.1 M solution). This step is essential, since, without a sufficient metal complexation inositol phosphates, in particular InsP_5 isomers and InsP_6, tend to co-precipitate with potassium perchlorate and, furthermore, are lost during charcoal treatment (see below). The amount of EDTA added should be in two-fold excess over the amount of bivalent cations present in the extracts (we assume 20 μmol of Mg^{2+} plus Ca^{2+} per g of wet cell or tissue weight).

In the case of cells in suspension, most of the bivalent cations are present in the medium. The minimal amount of EDTA should be 5 μmol.

Removal of Acid

To remove TCA, three to four extractions with equal volumes of ether are performed. Volumes of aqueous phases before and after ether extraction are determined. To remove PCA, an equimolar amount of 4 M KOH is added, with pH controlled by a pH electrode. The final pH in no case should exceed 6.0, since otherwise inositol phosphates tend to co-precipitate. The acetic acid added together with PCA (see above) facilitates the adjustment of the pH to between 5 and 6. The potassium perchlorate precipitate formed after 30 min on ice is removed by centrifugation at 5000 *g* for 5 min (in the cold). By weighing the wet and dry precipitates, the trapped volumes (usually about a third of the pellet weight) can be determined.

The remaining aqueous phases are subsequently freeze-dried. Volumes up to 2.5 ml are freeze-dried in 12 ml polypropylene tubes, larger volumes in 50 ml Falcon tubes. Tubes are covered with polypropylene stoppers or screw caps, respectively. By freezing the tubes almost horizontally, a thin layer forms which is more rapidly freeze-dried. Before freeze-drying, perforated Parafilm is substituted for the stoppers or caps (wear gloves when doing so!).

Charcoal Treatment

From physiological extracts nucleotides must be removed by charcoal treatment. This crucial step should be optimized for each type of sample by carefully following the effect of increasing amounts of charcoal or of repeated charcoal treatment on the disappearance of the peaks due to nucleotides (on the UV absorbance chromatograms) and on the recovery of inositol phosphates and internal standards. Therefore, only a starting protocol can be given. The samples should be taken up in as little water as possible in order to obtain high concentrations of salt during the charcoal treatment. As, to some degree, every batch of charcoal behaves like an anion exchanger, inositol phosphate recovery is thus increased. The second even more successful ingredient for optimal charcoal treatment is the cation chelator added after extraction (see above). Our explanation for the strongly improved recovery of inositol phosphates in the presence of a chelator even at low salt concentrations is that charcoal might contain hydroxyapatite-like cores which are "shielded" by the chelator. Boiling the charcoal prior to its use in HCl (see Chemicals, above) appears to reduce this undesirable property. With optimized charcoal additions, recoveries of internal InsP_5 or InsP_6 standards of close to 100% are possible.

Procedure

After the freeze-dried material has been dissolved in 0.5–2 ml of water and transferred into Eppendorf tubes of 1.3 or 2.5 ml capacity, the pH is checked again by pipetting 1 μl of the solution on to pH paper. When the pH is more than 6.0, it is adjusted to 6 by adding acetic acid. Unless nucleotides are also present in the cell suspension medium, the amount of charcoal added should be adjusted to the cell or tissue wet weight from which the sample is derived. To an equivalent of 100 mg of wet cell or tissue weight we add routinely 20 μl of a 20% (w/v) suspension of acid-treated Norit A in 0.1 M NaCl. For precise additions, the charcoal stock suspension should be continuously stirred and pipetted with truncated Eppendorf tips. The resulting suspension is kept on ice for 15 min and vortexed every 2 min. Then it is centrifuged at room temperature for 5 min at 5000 *g* in a bench-top centrifuge. The whole supernatant is carefully transferred to a second tube and once more treated with the same amount of charcoal. The two charcoal pellets are subsequently washed with 200 μl of 0.1 M NaCl and this wash is combined with the extract. In the case of muscle extracts, a third identical charcoal treatment has been found to be necessary to remove the high concentrations of adenine nucleotides present. Extracts of tissue specimens or sedimented cells do not contain much carrier salt and can be analysed directly after charcoal treatment. The charcoal fines are removed by another brief centrifugation and the samples are brought to 2.2 ml by adding 5 mM sodium acetate (pH 5), containing 1 mM NaF. When they are not to be analysed immediately, they are stored frozen at −20°C. Rethawed samples are vortexed and made dust free by centrifugation in a bench-top centrifuge. Volumes of 0.7, 1.2 or 2.15 ml are pipetted into appropriate autosampler bottles for injection with 0.5, 1.0 or 2.0 ml sample loops, respectively. When controls for acid-labile hexose phosphates or polyphosphates that interfere with inositol phosphate detection are to be made, they should be done after charcoal treatment by boiling the samples briefly in acid (e.g., 0.2 M HCl for 3, 7 and 20 min) and reneutralizing them with triethanolammonium. Otherwise multiply phosphorylated products arise as artefacts from the nucleotides.

Solid-phase Extraction

If samples contain much salt, e.g., when a large volume of cell suspension buffer is present initially, the salt must be removed prior to HPLC chromatography. Otherwise, peak retention times on anion exchangers are reduced and will tend to vary from sample to sample. Two variants of solid-phase extraction, both employing Q-Sepharose as the adsorbent and strong acid as the final eluant, are described below. We have also tried a solid-phase extraction of DEAE-Sepharose, employing ammonia as the inositol phosphate eluant (which acts by deprotonating the exchanger groups of the matrix) but the recoveries were insufficient.

Polypropylene columns (1 ml column size, self-made from Sarsted pipette tips) equipped with 10 ml buffer reservoirs (one-way syringe cylinders) are positioned in racks facilitating the simultaneous collection of effluents from 24 columns. Portions (0.5 g) of wet packed resin are used to fill the columns as a 1:1 slurry in H_2O. Samples are diluted up to 50 ml with H_2O and applied by gravity. The columns are subsequently washed with dilute acid (see below) without collection of the effluent. Finally the inositol phosphates are eluted with strong acid and the effluent collected in polypropylene tubes.

Method 1

Q-Sepharose is brought to the chloride form by washing with 20 volumes of 2 M NaCl, 10 volumes of 0.55 M HCl and then with H_2O. After sample application, columns are washed with 2 × 3 ml of 2 mM HCl. Inositol phosphates (Ins*P* to InsP_6) are subsequently eluted by 2 × 2 ml of 0.55 M HCl. The HCl is removed by freeze-drying from a thin layer (see above for the technique).

Method 2

Q-Sepharose is brought to the perchlorate form by washing with 20 volumes of 1 M sodium perchlorate, 10 volumes of 0.2 M PCA and finally water. After sample application, columns are washed with 2 × 10 ml of 0.8 mM PCA. Inositol phosphates (Ins*P* to InsP_6) are subsequently eluted by 2 × 1.7 ml of 0.15 M PCA. The acid is removed as potassium perchlorate by adding 50 μl of 2 M acetic acid, 50 μl of 0.1 M EDTA and then about 0.3 ml of 1.5 M KOH until the pH is 5.5. After the precipitate has been removed by centrifugation, the solution is freeze-dried, redissolved in 0.5 ml of H_2O and recentrifuged to remove the small precipitate of potassium perchlorate which has formed again.

For injection, the samples are dissolved or diluted, as necessary, in 5 mM sodium acetate (pH 5.5), containing 1 mM NaF. The presence of fluoride (see above) is especially important when the samples are standing in the autosampler at room temperature for longer periods.

Comments

Method 1 leads to salt-free samples, i.e., excellent retention time stabilities, but, due to a potential formation of azeotropic HCl at the end of the freeze-drying, some inositol phosphates exhibit a phosphate migration without a significant dephosphorylation after the treatment (17). When acid-labile phosphocompounds are to be analysed, method 2, which employs less strong acid and avoids the critical freeze-drying step, is to be recommended. However, its recoveries are somewhat lower than those of method 1.

CONTROLS

In order to confirm further the identity of inositol phosphate peaks, some controls can be performed from a set of identical charcoal-treated samples. We treat two of five identical samples with 1 M PCA at 100°C for 5 and 15 min, respectively, in order to remove acid-labile compounds that interfere with inositol phosphate peaks; we also treat one of the samples, adjusted to pH 8.0 with triethanolamine, with 10 units of alkaline phosphatase at 35°C for 16 h in order to remove most of the phosphomonoesters, including inositol phosphates. Inositol phosphates containing unclustered phosphate groups (i.e., lacking neighboring phosphate groups) are completely removed by the latter treatment. Treatment with 5-phosphatase from washed red cell membranes as described (25), removes only Ins(1,4,5)P_3 and Ins(1,3,4,5)P_4. The phosphatase-treated samples and an untreated control are extracted with PCA and the PCA is finally removed from all samples as described above.

Acknowledgments. The expert technical assistance by Friedhelm Vogel and Cornelia Tietz is greatly acknowledged. Thanks are due to Martina Nehls and Roger Falke for providing Figs. 3 and 4. This work was supported by grants from the Deutsche Forschungsgemeinschaft (Ma 989), from the Bundesminister für Forschung und Technologie, the Fonds der Chemischen Industrie, and by Boehringer Mannheim.

REFERENCES

1. Palmer S, Wakelam MJO. The Ins(1,4,5)P_3 binding site of bovine adrenocortical microsomes – function and regulation. *Biochem J* 1989;260:593–596.
2. Donie F, Reiser GA. A novel, specific binding protein assay for quantitation of intracellular inositol 1,3,4,5-tetrakisphosphate (InsP_4) using a high affinity InsP_4 receptor from cerebellum. *FEBS Lett* 1989;254:155–158.
3. Underwood RH, Greenley R, Glennon ET, Menachery AI, Braley LM, Williams GH. Mass determination of polyphosphoinositides and inositol trisphosphate in rat adrenal glomerulosa cells with a microspectrophotometric method. *Endocrinology* 1988;123:211–219.
4. Rittenhouse SE, Sasson JP. Mass changes in *myo*-inositol trisphosphate in human platelets stimulated by thrombin. *J Biol Chem* 1985;260:8657–8660.
5. MacGregor LC, Matschinsky FM. An enzymatic fluorimetric assay for *myo*-inositol. *Anal Biochem* 1984;141:382–389.
6. Irving GC, Cosgrove DJ. Interference by *myo*-inositol hexaphosphate in inorganic orthophosphate determinations. *Anal Biochem* 1970;36:381–388.
7. Meek JL, Inositol bis-, tris-, and tetrakis(phosphate)s: analysis in tissues by HPLC. *Proc Natl Acad Sci USA* 1987;83:4162–4166.
8. Haddad PR, Heckenberg AL. Determination of inorganic anions by high-performance liquid chromatography. *J Chromatogr* 1984;300:357–394.

9. Smith RE, MacQuarrie RA. Determination of inositol phosphates and other biologically important anions by ion chromatography. *Anal Biochem* 1988;170:308–315.
10. Franklin GO. Ion chromatography. *Int Laboratory* 1985;56–60
11. Smith RE, MacQuarrie RA, Jope RS. Determination of inositol phosphates and other anions in rat brains. *J Chromatogr Sci* 1989;27:491–495.
12. Rocklin RD, Pohl CA. Determination of carbohydrates using anion exchange chromatography with pulsed amperometric detection. *J Liquid Chromatogr* 1983;6:1577–1583.
13. Sherman WR, Ackerman KE, Berger RA, Gish BG, Zinbo M. Analysis of inositol mono- and polyphosphates by gas chromatography/mass spectrometry and fast atom bombardment. *Biomed Environ Mass Spectrom* 1986;13:333–341.
14. Hirvonen MR, Lihtamo H, Savolainen K. A gas chromatographic method for the determination of inositol monophosphates in rat brain. *Neurochem Res* 1988;13:957–962.
15. Mayr GW. A novel metal–dye detection system permits picomolar-range h.p.l.c. analysis of inositol polyphosphates from non-radioactively labelled cell or tissue specimens. *Biochem J* 1988;254:585–591.
16. Radenberg T, Scholz P, Bergmann G, Mayr GW. The quantitative spectrum of inositol phosphate metabolites in avian erythrocytes, analyzed by proton n.m.r. and h.p.l.c. with direct isomer detection. *Biochem J* 1989;264:323–333.
17. Pittet D, Schlegel W, Lew D, Monod A, Mayr GW. Mass changes of inositol tetrakis- and pentakisphosphate isomers induced by chemotactic peptide stimulation in HL-60 cells. *J Biol Chem* 1989;264:18,489–18,493.
18. Elchuk S, Cassidy RM. Separation of the lanthanides on high-efficiency bonded phases and conventional ion-exchange resins. *Anal Chem* 1979;51:1434–1438.
19. Koppitz B, Vogel F, Mayr GW. Mammalian aldolases are isomer-selective high-affinity inositol polyphosphate binders. *Eur J Biochem* 1986;161:421–433.
20. Irvine RF, Letcher AJ, Heslop JP, Berridge MJ. The inositol tris/tetrakisphosphate pathway – demonstration of Ins(1,4,5)P_3 3-kinase activity in animal tissues. *Nature* 1986;320: 631–634.
21. Phillippy BQ, White KD, Johnston MR, Tao S-H, Fox MRS. Preparation of inositol phosphates from sodium phytate by enzymatic and nonenzymatic hydrolysis. *Anal Biochem* 1987;162:115–121.
22. Mayr GW, Dietrich W. The only inositol tetrakisphosphate detectable in avian erythrocytes is the isomer lacking phosphate at position 3: a NMR study. *FEBS Lett* 1987;213:278–282.
23. Graf E, Empson KL, Eaton JW. Phytic acid, a natural antioxidant. *J Biol Chem* 1987;262:11,647–11,650.
24. Lanzetta PA, Alvarez LJ, Reinach PS, Candia OA. An improved assay for nanomole amounts of inorganic phosphate. *Anal Biochem* 1979;100:95–97.
25. Irvine RF, Letcher AJ, Lander DJ, Downes CP. Inositol trisphosphates in carbachol-stimulated rat parotid glands. *Biochem J* 1984;223:237–243.
26. Mayr GW. Picomol-mass HPLC analysis of inositol polyphosphates and other polyvalent oxyanions by a novel postcolumn derivatization technique: metal–dye detection (MDD). In: *Würzburger Chromatographic – Gespräche* (*Ultrochrom '88*), vol. 3. Freiburg: Pharmacia LKB GmbH, 1989;45–52.

Methods in Inositide Research,
edited by Robin F. Irvine.
Raven Press, Ltd., New York © 1990

8

Quantification of *myo*-Inositol Trisphosphate from Human Platelets

Susan E. Rittenhouse and Warren G. King

Department of Biochemistry, University of Vermont College of Medicine, Burlington, Vermont 05405

The methodology to be presented is based upon work described originally in 1985 (1) and amplified in 1987 (2). It is applicable to any *myo*-inositol-containing species, since ultimately *myo*-inositol, rather than *myo*-inositol phosphate, is quantified by gas-chromatographic (GC) analysis. The procedure is best suited to preparations containing relatively few contaminating sugars (or other substances that can be derivatized by trimethylsilylation and contribute other peaks to the GC chromatogram). *myo*-Inositol trisphosphates (InsP_3) can be separated relatively easily from such contaminants.

Our work focusses on human platelets, which are a relatively rich source of phosphoinositide polyphosphate. The most potent stimulus for phosphoinositidase C activation in these anucleate cells is thrombin, which leads to an increase of as much as 270 pmol/10^9 platelets in total InsP_3 in 30 s (or 650 pmol/10^9 platelets in 30 s when staurosporine is included to inhibit protein kinase C (3)).

EXTRACTION OF InsP_3 FROM PLATELETS

Human platelets (a minimum of 3 × 10^9/determination) are prepared in a physiological buffer (pH 7.3) (4). Platelet incubations are quenched at the appropriate time with 3.75 volumes of $CHCl_3$/MeOH/2 M HCl (2:5:0.25, by vol.). Any platelet aggregates present are dispersed thoroughly by bath sonication and samples are placed on ice for 15 min. Lipid extracts are processed fairly rapidly at this point, since (i) lipid extraction occurs virtually instantaneously and (ii) waiting increases the risk of hydrolysis of phosphatidylinositol 4,5-bisphosphate (PtdIns(4,5)P_2), that can contribute substantially to basal levels of inositol 1,4,5-trisphosphate (Ins(1,4,5)P_3). To monitor recoveries of inositol trisphosphates throughout the procedure, approximately 60,000 d.p.m. of [^{3}H]Ins(1,4,5)P_3 tracer (20 Ci/mmol or highest specific activity available) are added to each sample. Phases are split with one volume of $CHCl_3$ and 0.5 volume of 1 M HCl and samples are centrifuged at 1500 *g* at 4°C for 10 min. Aqueous upper phases are carefully removed with a siliconized pasteur pipette and transferred to clean siliconized tubes. The aqueous upper phases are backwashed with two volumes of $CHCl_3$ and vortexed vigorously. Extracts are again centrifuged as above and the aqueous upper phases are transferred to 50-ml tubes. Remaining chloroform in the samples is evaporated by passing nitrogen gas over them for 10 min.

PREPARATION OF InsP_3-RICH FRACTION

The samples are titrated to pH 7.0 with 1 M $NaHCO_3$ and diluted to 50 volumes with H_2O. InsP_3 is separated from other inositol phosphates by use of 0.7 cm internal diameter siliconized columns containing 1.0-ml bed volumes of BioRad Dowex AG 1-X8 (formate form), pre-washed with 10 ml of H_2O (5). Following pre-washing, the diluted upper phases obtained above are loaded on to the columns and allowed to flow through. Inositol phosphates (Ins*P*s) and InsP_2s are eluted with a 20 ml wash of 0.4 M ammonium formate, 0.1 M formic acid. InsP_3s are eluted and collected in 50-ml tubes with 10 ml of 0.8 M ammonium formate, 0.1 M formic acid. Samples are diluted to 50 ml with H_2O and desalted on 4.0 ml bed volumes of DEAE-Sephacel (Pharmacia) in 1.5-cm siliconized columns (6). Columns are pre-equilibrated with 10 ml of 1.0 M triethylamine ammonium bicarbonate (TEAB) (pH 7.8), followed by 10 ml of 0.1 M TEAB. The loaded columns are washed with 10 ml of 0.1 M TEAB followed by collection of InsP_3s in siliconized 13 mm × 100 mm tubes with 8.0 ml of 1.0 M TEAB. The 1.0 M fractions are freeze-dried to remove TEAB (3).

HYDROLYSIS OF InsP_3

Freeze-dried samples are dissolved in 200 μl of alkaline phosphatase buffer (5 mM $MgCl_2$, 19.8 mM ethanolamine (pH 9.5), 760 units alkaline phosphatase/ml (Sigma type VII, bovine intestinal enzyme)) and incubated for 2 h at 37°C (1). Note that this procedure will not lead to hydrolysis of glycerophosphoinositol 4,5-bisphosphate, since phosphatase does not hydrolyse the phosphodiester bond. Incubations are quenched with 10 μl of 0.1 M HCl. Samples are subsequently desalted on 0.5 ml columns of Rexyn mixed-bed ion exchange resin (Fisher Scientific, Springfield, NJ) in 0.7 cm internal diameter columns. Samples are applied to columns pre-washed with H_2O, and the eluate collected in 1.5-ml Eppendorf vials. Columns are further eluted with 600 μl of H_2O and the fractions are freeze-dried.

DERIVATIZATION OF SAMPLES FOR GAS CHROMATOGRAPHY

The dried samples are resuspended in 200 μl of H_2O and 10 μl are removed and counted to monitor recoveries. Samples are transferred to acid-washed 0.3-ml Reactivials (Pierce Chemical Co., Rockford, IL) containing 3.0 μl of 2.5 mM D-*chiro*-inositol as internal standard (1,2). To ensure integration of the *myo*-inositol peak during GC, 40 pmol of *myo*-inositol must be included in samples. The *myo*-inositol added to samples must be subtracted from final mass values. An individual standard of *chiro*-inositol/*myo*-inositol is prepared by adding 3 μl each of 2.5 mM *chiro*-inositol (7.5 nmol) and 1 mM *myo*-inositol (3 nmol) to a 0.3 ml Reactivial. To each sample and standard are added 50 μl of pyridine, followed by complete evaporation under nitrogen gas. To ensure complete drying of samples, 2 × 25 μl additions of pyridine are added to samples with evaporation under nitrogen gas. Residues are mixed with 25 μl of pyridine and 25 μl of Trisil concentrate (Pierce Chemical Co.), capped tightly, vortexed, and heated for 30 min at 70°C in a heater block, with vortexing every 5 min. Samples are then cooled to room temperature and centrifuged at 1500 *g* for 5 min. We have found this procedure to be >95% efficient for derivatization of standards (i.e. >95% [^{3}H]*myo*-inositol becomes hexane-soluble). Each investigator must verify that this is so in initial studies, since one assumes in subsequent calculations that all 3 nmol of *myo*-inositol have been derivatized as standard.

ANALYSIS BY GAS CHROMATOGRAPHY

Samples are resolved on a Hewlett-Packard Gas Chromatograph Model 5880A (1,2). Derivatives (3 μl) are injected in the split mode (56:1) at 250°C on to a 50 m fused silica WCOT CPSil5-CB capillary column (Chrompack). A split ratio of 56:1 minimizes solvent effects during the run and still gives acceptable sensitivity. Occasionally, sample derivatives must be concentrated under nitrogen gas to ensure integration of the *myo*-inositol peak, which is made feasible by the inclusion of the *chiro*-inositol standard, above. Separation of *chiro*- and *myo*-inositol is obtained under isothermal (200°C) conditions (run time, 25 min) with a He flow rate of approximately 2.27 cc/min. Make-up gas consists of nitrogen at 30 cc/ min. Masses are quantified by flame ionization detection (FID) at 250°C with 32 cc hydrogen/min and 400 cc air/min. Under these conditions *chiro*-inositol internal standard and *myo*-inositol yield retention times of about 14.2 min and 20.5 min, respectively. Peak area is determined by integration of the FID signal and final results are normalized with respect to recoveries of [^{3}H]Ins(1,4,5)P_3 tracer and *chiro*-inositol added as internal standard. We have determined that the linear range of this assay with a 56:1 split is 10 to at least 180 pmol of *myo*-inositol (added in 3 μl).

CALCULATION OF RESULTS

$$\textit{chiro}\text{-Inositol correction factor } (CCF) = \frac{\textit{chiro}\text{-inositol internal standard sample area}}{\textit{chiro}\text{-inositol standard area}}$$

$$\textit{myo}\text{-Inositol corrected area } (MCA) = \textit{myo}\text{-inositol sample area}/CCF$$

$$\textit{myo-Inositol(pmol)/unit area } (K) = (1/\textit{myo}\text{-inositol standard area}) \times y \text{ pmol of inositol standard}$$

Here y is 3000, since 3 nmol were derivatized initially. This equation provides a correction for the split ratio and injection volume.

$$[^3\text{H}]\text{Ins}(1,4,5)P_3 \text{ trace recovery factor } (RF) = \frac{20 \times \text{d.p.m. } [^3\text{H}]\text{Ins}(1,4,5)P_3 \text{ recovered}}{\text{d.p.m. } [^3\text{H}]\text{Ins}(1,4,5)P_3 \text{ added}}$$

The sampling factor is 20, since 5% was counted.

Finally,

$$\text{Ins}P_3 \text{ (pmol}/10^9 \text{ platelets)} = \frac{[(MCA \times K) - 40 \text{ pmol added } \textit{myo}\text{-inositol}]/RF}{\text{platelet no. per sample}/10^9}$$

Acknowledgment This work was supported by National Institutes of Health grant HL38622 (to S.E.R.).

REFERENCES

1. Rittenhouse SE, Sasson JP. Mass changes in *myo*-inositol trisphosphate in human platelets stimulated by thrombin: inhibitory effects of phorbol esters. *J Biol Chem* 1985;260:8657–8660.
2. Rittenhouse SE. Measurement by capillary gas chromatography of mass changes in *myo*-inositol trisphosphate. In: Conn PM, Means AR, eds. *Methods in enzymology: cellular regulators,*, Part B, *Calcium and lipids*, vol. 141. Orlando FL: Academic Press, 1987;143–149.
3. King WG, Rittenhouse SE. Inhibition of protein kinase C by staurosporine promotes elevated accumulations of inositol trisphosphates and tetrakisphosphates in human platelets exposed to thrombin. *J Biol Chem* 1989;264:6070–6074.
4. Rittenhouse SE. Activation of human platelet phospholipase C by ionophore A23187 is totally dependent upon cyclo-oxygenase products and ADP. *Biochem J* 1984;222:103–110.
5. Downes CP, Michell RH. The polyphosphoinositide phosphodiesterase of erythrocyte membranes. *Biochem J* 1981;198:133–140.
6. Tarver AP, King WG, Rittenhouse SE. Inositol 1,4,5-trisphosphate and inositol 1,2-cyclic 4,5-trisphosphate are minor components of total mass of inositol trisphosphate in thrombin-stimulated platelets. *J Biol Chem* 1987;262:17,268–17,271.

Methods in Inositide Research,
edited by Robin F. Irvine.
Raven Press, Ltd., New York © 1990

9

A Sensitive and Specific Mass Assay for *myo*-Inositol and Inositol Phosphates

John A. Maslanski and William B. Busa

Department of Biology, The Johns Hopkins University, Baltimore, Maryland 21218

A variety of techniques has been used to study the metabolism of the inositol phosphates. Techniques that have been employed to quantify inositol phosphate mass include phosphate detection or metal–dye binding following high-performance liquid chromatography (HPLC) separation (1,2), enzymatic oxidation (3,4), gas chromatography–mass spectrometry (5), bioluminescence (6), microspectroscopy (7), radiophosphorylation (8), and a radioreceptor assay (9). None of these mass assay methods, however, has gained wide acceptance, for reasons that include low sensitivity, lack of applicability to all inositol phosphates, or the requirement for advanced instrumentation. Thus, the most commonly employed technique has relied on the use of radioactively labelled substrates. While these radioactive tracer methods provide invaluable information on inositol phosphate fluxes, they do not usually permit their true intracellular levels to be determined.

A number of inherent problems arise when one attempts to quantify inositol phosphate levels using the radiotracer method. Substrates must be labelled to an isotopic steady-state if changes in the amount of labelled product are to reflect proportional changes in the actual mass of that product in the cell. For many cell systems this condition is difficult or impossible to meet due to problems such as poor uptake of label (our unpublished observations), *de novo* synthesis of substrates (10), metabolic compartmentation of substrates (11,12), or (e.g., in developing embryos) the non-steady state condition of the cells themselves. These difficulties in quantifying *myo*-inositol and its phosphates using the radiolabel technique can be appreciated best in those studies where mass determinations by other means have resulted in a discrepancy between the two values obtained (9). A further problem is one of cost and availability. Radiolabelled substrates for phosphoinositide research are expensive, and many are commercially unavailable.

We present here a mass assay for the determination of *myo*-inositol and

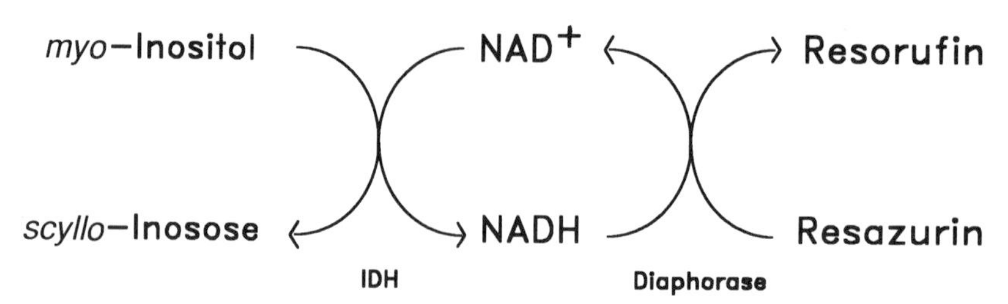

FIG. 1. A summary of the enzymatic reaction sequence for the assay presented. IDH, *myo*-inositol dehydrogenase.

inositol phosphates in physiological samples (13). The assay is based on two coupled enzyme reactions linked by the reduction and reoxidation of NAD (Fig. 1). *myo*-Inositol dehydrogenase (IDH, EC 1.1.1.18), in the presence of NAD, is used to oxidize free *myo*-inositol to *scyllo*-inosose and NADH (3). The NADH thus formed is then stoichiometrically reoxidized by the enzyme diaphorase. The electron acceptor in this second reaction is the non-fluorescent dye resazurin. Upon reduction, resazurin is converted to the intensely fluorescent compound resorufin, which is then quantified using a fluorometer. The principles behind the quantification of dehydrogenases using diaphorase and resazurin have been investigated (14–17).

The assay is highly sensitive (with a detection limit of about 10 pmol) and is specific for *myo*-inositol. Figure 2 shows a typical standard curve of free *myo*-inositol introduced at the IDH step. As can be seen, the assay is linear to at least 200 pmol. Inositol phosphates are determined by dephosphorylation using alkaline phosphatase (alternative methods of dephosphorylation may work as well) just prior to the reactions shown in Fig. 1. It takes approximately 4 h to complete the assay once the inositol phosphates are isolated, 1 h if free *myo*-inositol alone is to be determined. Of that time, less than 1 h is actual operator intervention for both the *myo*-inositol and inositol phosphate determinations. Thus, the present assay is sensitive, specific, simple, and inexpensive.

MATERIALS

IDH (highly purified grade) is, to our knowledge, available only from Sigma. The activity of the enzyme should be checked immediately upon receipt, since, in our hands, certain batches of IDH were not as active as stated and two were devoid of any activity. Assay conditions are provided by the manufacturer. Diaphorase (type II-L) and NAD (type VII, sodium salt) are also from Sigma. Hexokinase (ammonium sulphate suspension, from yeast) is from Boerhinger Mannheim Biochemicals. Alkaline phosphatase (high spe-

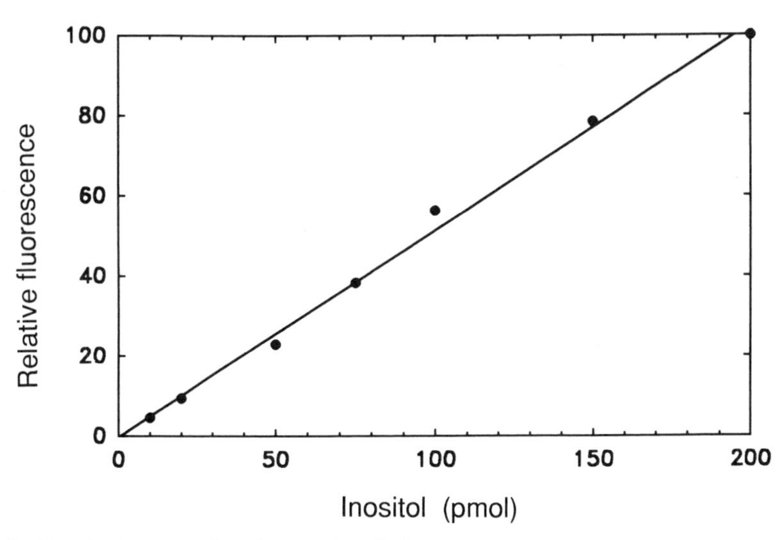

FIG. 2. Standard curve using the *myo*-inositol assay system. *myo*-Inositol standards were introduced at the IDH step. Each data point is a single determination, although these data represent a typical standard curve. Some day-to-day variability has been observed in absolute fluorescences.

cific activity molecular biology grade) is from Boerhinger Mannheim (the less expensive ammonium sulphate suspension is also usable).

Triethylammonium hydrogen carbonate (or triethylammonium bicarbonate, TEAB) is from Fluka (1.0 M solution, pH 8.4–8.6), and is diluted with deionized water to the concentrations specified in the text without adjusting the pH. Alternatively, TEAB may be made (at considerable savings!) from triethylamine and CO_2 as follows. Triethylamine (TEA, Fluka, *puriss* grade) is chilled on ice in a volumetric flash and diluted to 1.0 M by slow addition of ice-cold deionized water, with stirring. This solution is then bubbled, on ice, with CO_2 previously bubbled through deionized water. Sintered glass bubblers which produce very fine bubbles are best, as the end-point is reached in a manageable time with these. The pH of the TEA solution is continuously monitored with a temperature-compensated pH meter, and the solution is bubbled with CO_2 to a pH of 8.4. This endpoint should be approached slowly, as acidification will continue briefly after bubbling is terminated. Finally, the TEAB is filtered under vacuum (we use 0.45 μm nylon filters). A small amount of CO_2 gas will be evolved during this procedure, but the pH should not change significantly. The TEAB should be stored in tightly sealed brown-glass bottles at 4°C.

Resazurin is from Aldrich. Resazurin is anywhere from 90% to 99% pure, although even 99% purity is insufficient for this assay. The contaminant that causes the most trouble is resorufin, the product of the final reaction in our assay. To ensure the highest sensitivity, the background fluorescence must be as low as possible; therefore, resorufin contamination must be kept to a minimum. Fortunately, resazurin is easily separated from resorufin by thin-layer chromatography (TLC). To do this, resazurin is dissolved in chloroform/methanol (2:1, v/v) at a concentration of about 0.5 mg/ml. Approximately 400 μl is streaked on to a Silica Gel 60 TLC plate (20 cm × 20 cm) that has previously been in an 80°C oven for about 2 h. The streaked resazurin is allowed to dry completely. The plate is placed in a TLC tank containing 100 ml of chloroform/methanol (4:1) to which two drops of 1 M KOH have been added. The plate is developed and allowed to air-dry. Two bands will be evident, the purple band being resazurin and the pink band that runs just above it resorufin. The resazurin band is scraped off of the plate and placed in a centrifuge tube. Resazurin is eluted from the silica gel by adding 3 ml of 1.0 M phosphate buffer (pH 6.5) to the tube, vortexing vigorously, and centrifuging at 1500 *g* for 5 min. The supernatant is withdrawn and the elution process is repeated, combining both supernatants. The purified resazurin is then put through a 0.45 μm syringe filter to remove any silica gel that was decanted with the resazurin. Quantification of the resazurin is accomplished by reading the absorbance from a spectrophotometer at 600 nm using an extinction coefficient of 44,640 l/mol per cm. For maximum stability, the resazurin should be kept at a concentration of at least 20 μM. In this state, the resazurin is stable for at least one month in room light, but breaks down when exposed to direct sunlight. It is usually safest to store it at 4°C in a refrigerator. Since the sensitivity of the assay depends on the background fluorescence, how long a preparation of resazurin remains usable depends upon the sensitivity requirements of your application. If maximum sensitivity is required at all times, then a preparation of resazurin will be good for only about two weeks. Otherwise, a preparation may last several months.

TISSUE PREPARATION

Although other methods of isolating *myo*-inositol or its phosphates may work just as well, the following method has been studied exhaustively and is well suited for work with the assay. Whatever method is used, care must be taken to use only plasticware, especially when one is assaying for inositol phosphates. We have found that the inositol phosphates will stick readily to glass. This may not be a problem if the sample contains other highly charged compounds in a greater abundance than the inositol phosphates, but in normal practice we use only plastic tubes and pipette tips.

The tissue to be assayed is placed in a 4 ml conical polypropylene tube (Sarstedt no. 57.512) and quickly frozen in a solid CO_2/methanol bath. The tissue is then homogenized, on ice, in 100 μl of 7.5% (w/v) ice-cold perchloric acid using a Teflon pestle. Wheaton offers a Teflon pestle (Wheaton no.

358133) that fits nicely into the bottom of these conical tubes. The pestle is rinsed with 100 μl of the perchloric acid and added to the homogenate. The homogenates are then kept on ice for approximately 20 min. The precipitate that forms is pelleted in a Sorval centrifuge at 10,000 *g*, at 4°C for 10 min. The supernatant is withdrawn and placed in a 4 ml conical tube. Each supernatant then receives 50 μl of 10 mM EDTA (pH 7.0 with KOH). For the subsequent steps, it is important that the samples remain on ice. Perchloric acid is removed from solution (as the K^+ salt) by the addition of ice-cold 10 M KOH until the pH is greater than 7.0. The samples are then left on ice for about 20 min. The potassium perchlorate is spun down at 2000 *g*, at 40°C for 10 min. Any water added to the tubes to balance the centrifuge should be ice cold and should be kept to a minimum. The supernatants are withdrawn and the pH of each is brought to between 6 and 9 by HCl. The volume of the extracts is not critical at this stage and, since it is easier to measure the pH of a larger volume than a smaller one, deionized water may be added.

CHROMATOGRAPHY

If only the free *myo*-inositol levels are to be determined, the samples are put through a column containing about 0.5 ml of a mixed-bed ion exchange resin (Dowex -50W 200–400 mesh). One ml plastic pipette tips work quite well as the columns. The resin beads are sufficiently large that they will be held up at the bottom of the pipette tip. The sample is added to the minicolumn, and the eluate is collected. The column is washed with about five column volumes of deionized water, again the eluate being collected. This procedure usually takes no more than about 10 min per sample and the total time may be decreased significantly by using multiple columns. The samples are then dried overnight in a vacuum centrifuge (Speed-Vac, Savant).

For the separation of inositol phosphates, we use Sep-Pak Accell Plus QMA cartridges (Waters, Millipore Corp., Milford, MA) instead of Dowex columns (13). The Sep-Pak columns require a smaller volume of eluant and can be run at a higher flow rate than the Dowex columns. This greatly reduces the time spent preparing the inositol phosphates for the assay. Additionally, the eluant employed is readily removed via vacuum centrifugation. Wreggett et al. (18) had established conditions whereby inositol phosphates could be separated on Sep-Pak cartridges, but then retracted the method when the manufacturer changed the resin (see advertisement facing p. 626 in *Nature* (1989), vol. 336). We have achieved separation of the inositol phosphates with the new Sep-Pak resin using TEAB as the eluant.

We usually employ a manifold set-up, manufactured by Analtech or other suppliers, which uses vacuum to draw the eluant through the columns. These work quite well when a large number of samples need to be chromatographed simultaneously.

Before first use, columns are washed with 10 ml of deionized water, 10 ml of 1.0 M TEAB (flow rate less than 0.5 ml/min) and 10 ml more of deionized water. Neutralized extracts are diluted to 5–10 ml with deionized water and applied to the washed Sep-Pak columns. Care should be taken not to allow the columns to run dry. After the samples are applied, the columns are washed with 4 ml of 0.02 M TEAB. The inositol phosphates are eluted from the column by 4 ml each of 0.1, 0.3, 0.4 M, and 6 ml of 0.5 M TEAB to elute the inositol mono-, bis-, tris- and tetrakisphosphates, respectively, at a flow rate not exceeding 2 ml/min. The sample application eluate and 0.02 M wash may be dried down and used to determine free *myo*-inositol levels (see above). It is not necessary to separate glycerophosphoinositol from inositol monosphosphate because in the subsequent dephosphorylation, alkaline phosphatase will be unreactive toward the phosphodiester. Each fraction is collected into 4 ml conical polypropylene tubes. The samples are then dried down via vacuum centrifugation. Before reuse, the columns are washed with 4 ml of 1.0 M TEAB followed by 10 ml of deionized water. Once the fractions have dried down, internal standards may be added to some of the samples (''spiking'', see Analysis of Results, below).

The above procedure is sufficient to elute free *myo*-inositol and inositol phosphates (from inositol monophosphates through inositol trisphosphates). Higher phosphorylated forms have not been investigated. Also, this procedure does not resolve specific isomers of the inositol phosphates. In order to analyse different isomers, an alternative method would be required (see Chapters 2, 5 and 7). In choosing an alternative separation scheme, it should be borne in mind that the following buffer systems are incompatible with the analysis of inositol phosphates using this assay: any eluant which is not reasonably volatile (i.e., cannot be easily removed by vacuum evaporation) and any eluant containing phosphate. Phosphate buffers cannot be used because the first step in the assay, the dephosphorylation of the inositol phosphates by alkaline phosphatase, would be inhibited by the high concentrations of phosphate. Ammonium-containing buffers, if possible, should be avoided. It seems that the small amount of ammonium that remains after dry-down is in some way inhibitory to one of the subsequent enzymatic reactions. Homogenization in trichloroacetic acid (TCA), likewise, should be avoided for similar reasons. Lowry and Passonneau (19) have noted that some dehydrogenases are particularly sensitive to TCA and to ammonium-containing compounds.

These incompatibilities largely rule out the use of the two most common techniques for separation of inositol phosphates (namely elution from Dowex columns with ammonium formate and HPLC with phosphate-containing buffers), although the technique described by Spencer et al. (Chapter 4), whereby inositol phosphates are eluted in HCl, might be suitable for use in conjunction with this mass assay. The Sep-Pak technique just described can substitute for Dowex (formate) chromatography, and the ion chromatographic separation technique of Sun et al. (Chapter 11) may substitute for phosphate-based HPLC techniques. Alternatively, one of the methods listed above may be used to effect a buffer exchange, replacing the incompatible buffer with TEAB. A final way around this problem may be to desalt the inositol on a mixed-bed ion exchange resin immediately before the addition of the IDH reagent.

ASSAY REAGENTS

Alkaline Phosphatase Reagent

If the high specific activity preparation is to be used, it is added directly to 0.1 M Tris·HCl (pH 9.0), 0.1 mM $ZnCl_2$ at a concentration of 200 units/ml. If the ammonium sulphate suspension is to be used, the precipitate must first be spun down and decanted. It is then resuspended in a minimum volume of 20 mM imidazole, 0.02% (w/v) bovine serum albumin (BSA) (pH 7.0) before addition to a Tris·HCl/$ZnCl_2$ (pH 9.0) solution such that the final concentration of enzyme is 200 units/ml and the Tris·HCl and $ZnCl_2$ concentrations are 0.1 M and 0.1 mM, respectively. The alkaline phosphatase reagent is made fresh each day.

Hexokinase Reagent

The ammonium sulphate precipitate is centrifuged and the salt solution is removed. The enzyme is resuspended in a solution of 50 mM Tris, 10 mM $MgCl_2$, 0.1 M ATP, 0.02% BSA (pH 9.0) at an enzyme concentration of 200 units/ml.

IDH Reagent

IDH is reconstituted in 10 mM phosphate, 0.02% BSA (pH 6.8) at a concentration of approximately 5 units/ml. It is stored at −80°C in small portions (0.1–0.2 units/portion). When needed, a portion is diluted to a concentration of 1 unit/ml with 10 mM phosphate (pH 6.8), 0.02% BSA. Once thawed, the enzyme is discarded at the end of each day.

Diaphorase Reagent

Diaphorase is reconstituted in 20 mM phosphate, 0.02% BSA (pH 6.8) at a concentration of 10 units/ml. It is stored in small portions at −80°C. As with the IDH reagent, it must be discarded at the end of each day.

NAD

The sodium salt of NAD is dissolved in deionized water at a concentration of 0.1 M. The pH of this solution will be approximately 9.0. Because NAD is unstable in alkaline solutions, it should be made just prior to use and kept on ice.

All enzyme reagents and diluents should be kept on ice at all times. Also, all solutions containing Tris are made using Tris Base, adjusting the pH with HCl.

ASSAY METHODS

Dephosphorylation of the inositol phosphates is accomplished by the addition of 45 μl of alkaline phosphatase reagent to the dried samples. The samples are incubated for 2 h at 37°C. The alkaline phosphatase is then inactivated by placing the samples in a water-bath at 100°C for 4 min.

If it is necessary to remove any contaminating traces of glucose (see below), 5 μl of hexokinase reagent is next added to each tube and the samples are then incubated at 37°C for 1 h. The hexokinase reaction is stopped by placing the samples in a water-bath at 100°C for 3 min.

Oxidation of the *myo*-inositol is carried out by the addition of 5 μl of 100 mM NAD and 5 μl of IDH reagent to each sample. The samples are incubated at room temperature for 15 min, after which 5 μl of 0.8 M HCl is added to decrease the pH to approximately 6.5. Each time HCl or Tris buffer is prepared, it should be checked by mixing appropriate amounts of Tris and HCl, then measuring the pH. If the pH of the resulting solution is not within 0.5 pH units of 6.5, then the HCl should be adjusted accordingly.

The quantification of NADH produced in the previous reaction is a modification of the procedure originally published by Guilbault and Kramer (14). Ten μl of 20 μM resazurin in 1.0 M phosphate (pH 6.5), and 5 μl of diaphorase, are added to each tube. If the resazurin was stored at a concentration higher than 20 μM, it should be diluted to 20 μM with 1 M phosphate (pH 6.5) just prior to its addition to the assay. The samples are incubated at room temperature in a darkened area (e.g., inside a cabinet) for 15 min, at which time 920 μl of 0.1 M Tris·HCl (pH 9.0) is added.

The tubes are thoroughly vortexed and each sample is transferred to an acrylic fluorometer cuvette. Resorufin fluorescence is measured on a fluorometer with the excitation and emission wavelengths set at 565 and 585 nm, respectively. Suitable bandwidths on our instrument are 10 nm for both the excitation and emission slits.

ASSAY SUMMARY

1. *Dephosphorylation:* To dried samples, add 45 μl of alkaline phosphatase solution. Incubate for 2 h at 37°C. Inactivate enzyme by placing samples in a water-bath at 100°C for 3 min. Cool tubes to at least 37°C.

2. *Hexose removal* (optional): Add 5 μl of hexokinase solution to each sample. Incubate for 1 h at 37°C. Place samples in a water-bath at 100°C for 3 min. Allow tubes to cool to room temperature.
3. *Inositol oxidation:* Add 5 μl each of the NAD and IDH solutions. Incubate at room temperature for 15 min. Decrease pH to approximately 6.5 by adding 5 μl of 0.8 M HCl.
4. *Resazurin reduction:* Add 10 μl of the resazurin solution and 5 μl of the diaphorase solution to each sample. Incubate at room temperature in a darkened area for 15 min.
5. *Resorufin measurement:* Add 920 μl 0.1 M Tris·HCl (pH 9.0) to each sample and transfer to fluorometer cuvette. Measure fluorescence at 565 nm (excitation) and 585 nm (emission).

ANALYSIS OF RESULTS

Besides samples, tissue blanks and internal standards ("spikes") also have to be assayed. The tissue blank is treated in a manner similar to that of the samples, except that it does not receive IDH. Instead, a solution of 10 mM phosphate, 0.02% BSA (pH 6.8) is substituted. Spiked samples are also assayed in order to generate an internal calibration (see below). We generally spike samples with a standard amount of *myo*-inositol or the inositol phosphate(s) to be assayed. One method of generating tissue blanks and spikes involves resuspending the dried-down fractions (after chromatography) in a small volume of deionized water and dividing each into three tubes. To one of these is added the appropriate inositol phosphate. The samples are then redried and assayed as described below. Alternatively, if it has been determined that all tissue samples of a particular fraction (e.g., the inositol monophosphate fraction) respond similarly with respect to the tissue blanks and sample spikes, then only one blank and one spike need be assayed for each fraction. These may arise from independent homogenates from typical tissue samples. Because IDH may contribute slightly to background fluorescence, two additional tubes must be included when tissue samples are assayed. One tube contains all assay reagents but no tissue. The other is treated identically to the tissue blank (see above), except that it also does not receive tissue. Subtracting the fluorescence of the second tube from that of the first yields the fluorescence due solely to IDH.

The calculation of the picomoles in a sample are determined by the following equation:

$$\text{pmol in sample} = (\text{sample RFU} - \text{blank RFU}) \times \frac{\text{pmol added in spike}}{(\text{spike RFU} - \text{sample RFU})}$$

where RFU stands for relative fluorescence units and blank is equal to the fluorescence of the tissue blank (tissue sample that does not contain IDH)

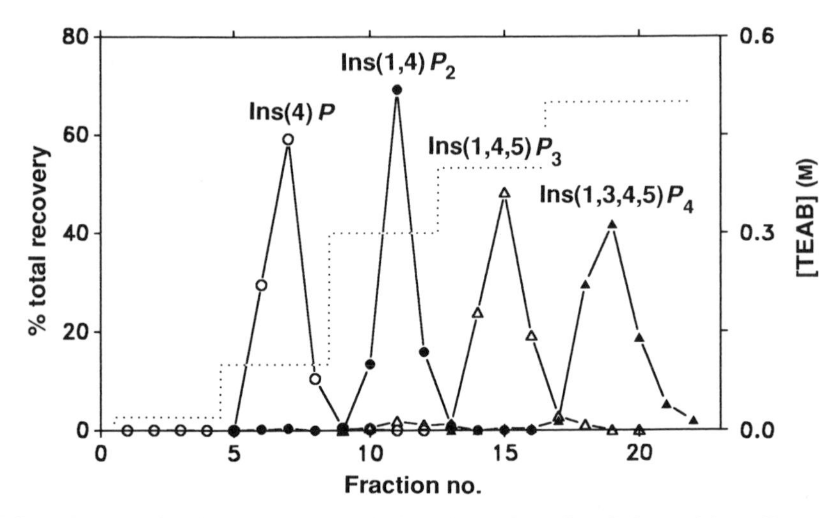

FIG. 3. Elution of radiolabelled inositol phosphates from Sep-Pak cartridges. Four separate columns were loaded with approximately 10,000 c.p.m. of either Ins(4)P, Ins(1,4)P_2, Ins(1,4,5)P_3, or Ins(1,3,4,5)P_4 in the presence of a rat liver extract (equivalent to 10 mg wet weight). The columns were then washed with 4 ml of 0.02 M TEAB. Elution of the inositol phosphates was accomplished by the sequential addition of 4 ml of 0.1, 0.3, 0.4 M, and 6 ml of 0.5 M TEAB to elute ins(4)P, ins(1,4)P_2, Ins(1,4,5)P_3, and Ins(1,3,4,5)P_4, respectively. Column flow rate was between 1 and 2 ml/min. The dotted line indicates the concentration of TEAB used to elute each of the inositol phosphates. For abbreviations, see the text.

plus the fluorescence of the IDH blank (the tube that contains IDH but no tissue) minus the fluorescence of resazurin (the tube that contains neither tissue nor IDH).

RESULTS AND DISCUSSION

Sep-Pak Chromatography

Recovery of authentic standards of either free *myo*-inositol or inositol phosphates from Sep-Pak columns is typically between 90% and 100%. Recoveries were assessed by monitoring radiolabelled standards and by the present mass assay. Fig. 3 shows a typical elution profile from a Sep-Pak column. The column was loaded with 5000 c.p.m. each of tritiated *myo*-inositol, inositol 4-monophosphate (Ins(4)P), 1,4-bisphosphate (Ins(1,4)P_2) and 1,4,5-trisphosphate (Ins(1,4,5)P_3) in a tissue extract equivalent to 10 frog oocytes (about 10 mg of tissue).

The columns may be reused at least 10–15 times. Resolution should be checked periodically by running radioactive standards through a typical elution protocol. When running radioactive standards, we generally include non-radioactive standards to increase reproducibility. Wreggett et al. (20) have noted that the inclusion of a phytic acid hydrolysate increases reproducibility and recovery of radiolabelled inositol phosphates from chromatography columns. Alternatively, radioactive standards may be chromatographed in the presence of a tissue extract to mimic actual operating conditions. Column degradation will first manfest itself as a loss of resolution between the InsP_2 and InsP_3 fractions. Also, there is some variability between batches of Sep-Pak cartridges. The variability is seen mostly in the resolution between the InsP_2 and InsP_3 fractions.

Specificity of the Assay

While the conditions presented above are sufficient fully to dephosphorylate 97% of Ins(1,4,5)P_3 (data not shown), Ins(1,3,4,5)P_4 is quite resistant to dephosphorylation by alkaline phosphatase (Maslanski and Busa, unpublished observations). Shayman et al. (4) have observed lower rates of hydrolysis of InsP_3 than expected and suggest the inefficiency of alkaline phosphatase in removing vicinal phosphates as the cause. This suggestion is supported by our observations that dephosphorylation using the conditions presented above will dephosphorylate less than one quarter of the added Ins(1,3,4,5)P_4, the remainder eluting with InsP_4 after chromatographic separation (data not shown). Quantification of InsP_4 under the present conditions may be accomplished if internal standards are used (see Standards, below), although at a lower sensitivity. Alternatively, one may use other methods of dephosphorylation of InsP_4 such as acid digestion. Again, inclusion of an internal standard will allow accurate quantification.

IDH displays low but potentially significant levels of activity with inositol isomers other than the *myo*-isomer (6). These minor interferences would not be expected to affect determinations of inositol phosphates, since, to our knowledge, phosphates of isomers other than *myo*-inositol have never been observed. Determinations of free *myo*-inositol levels could, in theory, be compromised, but existing data suggest that such effects would be negligible. For example, in rat brain, levels of *scyllo*- and *epi*-inositol are 50 and 1500-fold lower, respectively, than *myo*-inositol levels (21). The reactivity of IDH toward *scyllo*- and *epi*-inositol is 1.6% and 8%, respectively, of its reactivity toward *myo*-inositol (6). Thus, free inositol assays of this tissue would reflect levels of the *myo*-isomer only.

MacGregor and Matschinsky (3) have shown that a number of sugars and sugar alcohols do not react with IDH. One exception was glucose. Weissbach (22) has observed that IDH reacts with glucose, although at a greatly reduced rate compared with its reaction with *myo*-inositol. This may be problematic when trying to measure inositol levels in tissues where free glucose levels are high (e.g., in serum). Also, it is likely that phosphorylated forms of glucose (mostly in the form of glucose-6-phosphate) will co-elute with one or more of the inositol phosphates in the elution protocol outlined above. Free glucose will then be produced during the subsequent dephosphorylation. To remedy this, samples are treated with hexokinase. Hexokinase phosphorylates glucose, rendering it insensitive to IDH, while *myo*-inositol does not become phosphorylated. If it has been determined that the particular separation protocol used excludes glucose or glucose phosphate contamination, then the hexokinase step may be deleted.

Standards

For the purpose of calculations (see Analysis of Results, above), we generally assay sample extracts that have been spiked with an internal standard. Internal standards provide the most reliable method of assessing the efficiency of the enzymatic reactions that follow the chromatography and subsequent dry-down. In some cases, we have observed lower values than expected when standards were assayed in the presence of tissue extracts. This interference is greatest in the Ins$P$ fraction. Interference has been observed in the free *myo*-inositol and InsP_2 fractions, albeit at a reduced level, and it has not been observed in the InsP_3 fractions. The nature of this interference has not yet been investigated. The amount of the spike required is dependent on the extent of the interference. In the worst case, we have had to add 400 pmol of Ins(1)P, although for InsP_3, in which no interference is seen, we generally spike with 50 pmol of Ins(1,4,5)P_3. An external standard curve may be generated by introducing *myo*-inositol standards at the IDH step. These external standards merely ensure linearity in the working range of the samples. Small amounts of radioactive inositol phosphates may be added just after homogenization, which would allow recoveries for the entire procedure to be calculated.

Comparison with Other Techniques

The present assay has been used to quantify *myo*-inositol levels in the liver and testes of the rat. The rat was killed by cervical dislocation and the tissues were removed and placed into liquid nitrogen within 4 min of death. Table 1 compares our results with those of previous works [(23) and works cited therein)]. It is apparent that our results agree quite well with those that have been previously published.

This assay was also used to quantify inositol phosphates in the rat liver. We obtained values of 39.1 ($\pm$2.6), 6.97 ($\pm$0.28) and 3.33 ($\pm$0.10) nmol/g of tissue for Ins$P$, InsP_2, *Ins*P_3, respectively (mean $\pm$ S.E.M. of four to six replicates). Meek (1), using a phosphate detection scheme in conjunction with HPLC, has obtained values of 44 ($\pm$7) and 13 ($\pm$1) nmol/g for InsP_2

TABLE 1. *Comparison of free inositol levels in rat liver and testis*

Tissue	Present assay (*n*)	Other methods (*n*)[a]
Liver	0.97 ± 0.03 (2)	0.67 ± 0.19 (30)[b]
Testes	0.83 ± 0.09 (6)	1.60 ± 0.39 (9)[c]

Results using the present assay are expressed as replicates (*n*) of the same animal, while the results of Other methods summarize those from a number of different animals. Values shown are expressed as μmol/g tissue ± S.E.M.

[a] Dawson and Freinkel (23).

[b] Range = 0.33–1.36.

[c] Range = 0.78–1.83.

and InsP_3, respectively. It should be noted that the value for the InsP_3 reported by Meek (1) is for the 1,4,5 isomer only. While these values differ significantly from our results, a number of factors differed between our study and Meek's which may account for this. These include the method of sacrifice, which Meek demonstrated can have a significant influence on inositol phosphate levels, and the rapidity of tissue harvesting. Thus, the present data merely serve to demonstrate that the assay is adequately sensitive to measure "resting" inositol phosphate levels in tissues.

In a more direct comparison with an independent technique, we (13) have compared free *myo*-inositol levels in frog oocytes as determined either by the present assay or by gas chromatography–mass spectrometry. The values obtained, 31.6 (±4.9) and 30.9 (±1.3) pmol/oocyte, respectively, underscore the validity of the present assay.

In summary, we have presented here an assay for the mass determination of inositol. This assay is specific for the *myo*-isomer of inositol and can be used to determine the levels of either free *myo*-inositol or any inositol phosphate. It is highly sensitive and convenient, requiring very little operator time and only common laboratory instrumentation. By being able to determine intracellular concentrations of various products of inositol metabolism, one may obtain a better understanding of their complex interconversions and their importance to cellular physiology.

Acknowledgment The studies reported here were supported by National Institutes of Health grant HD22879.

REFERENCES

1. Meek JL. Inositol bis-, tris-, and tetrakis (phosphate)s: analysis in tissues by HPLC. *Proc Natl Acad Sci USA* 1986;83:4162–4166.
2. Mayr GW. A novel metal-dye detection system permits picomolar range h.p.l.c. analysis of inositol polyphosphates from non-radioactively labelled cell or tissue specimens. *Biochem J* 1988;254:585–591.

3. MacGregor LC, Matschinsky FM. An enzymatic fluorimetric assay for *myo*-inositol. *Anal Biochem* 1984;141:382–389.
4. Shayman JA, Morrison AR, Lowry OH. Enzymatic fluorometric assay for *myo*-inositol trisphosphate. *Anal Biochem* 1987;162:562–568.
5. Rittenhouse SE, Sasson JP. Mass changes in myoinositol trisphosphate in human platelets stimulated by thrombin. *J Biol Chem* 1985;260:8657–8660.
6. Gudermann TW, Cooper TG. A sensitive bioluminescence assay for *myo*-inositol. *Anal Biochem* 1986;158:59–63.
7. Underwood RH, Greely R, Eileen TG, Menachery AI, Braley LM, Williams GH. Mass determination of polyphosphoinositides and inositol triphosphate in rat adrenal glomerulosa cells with a microspectrophotometric method. *Endocrinology* 1988;123:211–219.
8. Tarver AP, King WG, Rittenhouse SE. Inositol 1,4,5-trisphosphate and inositol 1,2-cyclic 4,5-trisphosphate are minor components of total mass of inositol trisphosphate in thrombin-stimulated platelets: rapid formation of inositol 1,3,4-trisphosphate. *J Biol Chem* 1987;262:17,286–17,271.
9. Challiss RAJ, Batty IR, Nahorski SR. Mass measurements of inositol(1,4,5)trisphosphate in rat cerebral cortex slices using a radioreceptor assay: effects of neurotransmitters and depolarization. *Biochem Biophys Res Commun* 1988;157:684–691.
10. Horstman DA, Takemura H, Putney JW Jr. Formation and metabolism of [^{3}H]inositol phosphates in AR442J pancreatoma cells. *J Biol Chem* 1988;263:15,297–15,303.
11. King CE, Stephens LJ, Hawkins PT, Guy GR, Michell RH. Multiple metabolic pools of phosphoinositides and phosphatidate in human erythrocytes incubated in a medium that permits rapid transmembrane exchange of phosphate. *Biochem J* 1987;244:209–217.
12. Koreh K, Monaco ME. The relationship of hormone sensitive and hormone-insensitive phosphatidylinositol to phophatidylinositol 4,5-bisphosphate in the wrk-1 cell. *J Biol Chem* 1986;261:88–91.
13. Maslanski JM and Busa WB. Manuscript in preparation.
14. Guilbault GG, Kramer DN. New direct fluorometric method for measuring dehydrogenase activity. *Anal Chem* 1964;36:2497–2498.
15. Guilbault GG, Kramer DN, Fluorometric procedure for measuring the activity of dehydrogenases. *Anal Chem* 1965;37:1219–1221.
16. DeJong DW, Woodlief WG. Fluorometric assay of tobacco leaf dehydrogenases with resazurin. *Biochim Biophys Acta* 1977;484:249–259.
17. Barnes S, Spenney JG, Stoichiometry of the NADH-oxidoreductase reaction for dehydrogenase determinations. *Clin Chim Acta* 1980;107:149–154.
18. Wreggett KA, Irvine RF. A rapid separation method for inositol phosphates and their isomers. *Biochem J* 1987;245:655–660.
19. Lowry OH, Passonneau JV. *A flexible system of enzymatic analysis*. New York: Academic Press, 1972.
20. Wreggett KA, Howe LR, Moore JP, Irvine RF. Extraction and recovery of inositol phosphates from tissues. *Biochem J* 1987;245:933–934.
21. Sherman WR, Hipps PP, Mauck LA, Rasheed A. Studies on enzymes of inositol metabolism. In: Wells WW, Eisenberg F Jr, eds. *Cyclitols and phosphoinositides*. New York: Academic Press, 1978;279–295.
22. Weissbach A. The determination of myo-inositol. *Biochim Biophys Acta* 1958;27:608–611.
23. Dawson RMC, Freinkel N. The distribution of free *meso*inositol in mammalian tissues, including some observations on the lactating rat. *Biochem J* 1961;78:606–610.

Methods in Inositide Research,
edited by Robin F. Irvine.
Raven Press, Ltd., New York © 1990

10

Mass Measurement of Inositol 1,4,5-Trisphosphate Using a Specific Binding Assay

Susan Palmer and Michael J. O. Wakelam

Molecular Pharmacology Group, Department of Biochemistry, University of Glasgow, Glasgow G12 8QQ, UK

The method described is for the mass measurement of inositol 1,4,5-trisphosphate (Ins(1,4,5)P_3) (1). The sensitivity of the assay is such that 0.2 pmol of Ins(1,4,5)P_3 can be detected. In essence, the Ins(1,4,5)P_3 present in a cell extract competes with a fixed quantity of high specific activity [^{3}H]Ins(1,4,5)P_3 (although [^{32}P]Ins(1,4,5)P_3 could be used instead) for the Ins(1,4,5)P_3-specific binding sites of bovine adrenocortical microsomes. A standard curve using known amounts of unlabelled Ins(1,4,5)P_3 is conducted in parallel. Thus, the quantity of Ins(1,4,5)P_3 in the cell extract can be calculated.

PREPARATION OF BOVINE ADRENOCORTICAL MICROSOMES

Typically, approximately 20 bovine adrenal glands (150 g of tissue after dissection) will yield 80 ml of binding protein, which is sufficient for 3200 incubations. Adrenal glands should be obtained fresh if possible, although tissue that has been frozen at −20°C shortly after removal from the animal is also suitable. All tissue and buffers should be kept on ice throughout the preparation.

Each adrenal gland is cleaned of fat and bisected using scissors. The medulla is removed by scraping with a scalpel. The tissue is weighed and two volumes of ice-cold homogenization buffer (20 mM $NaHCO_3$, 1 mM dithiothreitol (pH 7.5)) added, e.g., 150 g of dissected tissue and 300 ml of homogenization buffer. The tissue is finely chopped in a Waring blender for approximately 1 min. Filtration of the tissue, through a 0.8 cm wire mesh, is preferable at this stage but not essential. The chopped tissue is then homogenized using a Potter-Elvehjem-type homogenizer.

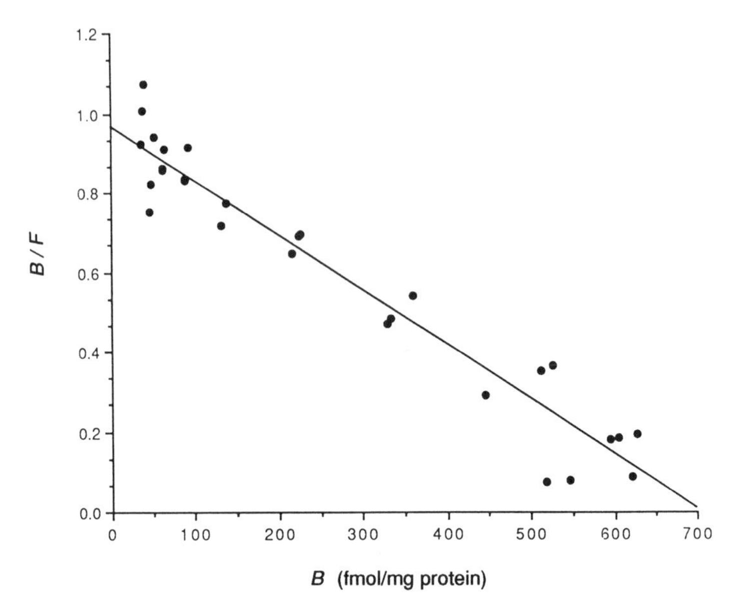

FIG. 1. Scatchard analysis of binding data. Results are from a single typical experiment. *B* is bound ligand and *F* is a free ligand.

The homogenate is centrifuged (Beckman J2-21 centrifuge, 8 × 50 ml rotor) at 5000 *g* for 15 min at 4°C. The resulting supernatant is centrifuged at 35,000 *g* for 20 min at 40°C. The pellet is resuspended in homogenization buffer and centrifuged again at 35,000 *g* for 20 min at 4°C. The final pellet is resuspended in homogenization buffer at 20 to 40 mg protein/ml.

For a preparation of this size, dissection takes approximately 2 h, blending and homogenization approximately 90 min and centrifugation etc. approximately the same: thus, 5 h are required for the complete preparation.

The binding protein, stored in 10 ml portions at −80°C, is stable for at least six months and can tolerate approximately four freeze-thaw cycles without loss of binding activity. For daily use of the binding protein, 1 ml portions, each of which is sufficient for 40 incubations, are stored.

Each preparation of the binding protein is characterized using the binding assay described below. Scatchard analysis of the binding data suggests the presence of a single population of binding sites (Fig. 1).

THE BINDING ASSAY

The binding assay is conducted in 2 ml polypropylene centrifuge tubes, obtained from Sarstedt. Each incubation contains: 25 μl of incubation buffer, 25 μl of [^{3}H]Ins(1,4,5)P_3, 25 μl of standard Ins(1,4,5)P_3 or cell extract, and finally 25 μl of binding protein.

1. Incubation buffer is: 100 mM Tris·HCl (pH 9), 4 mM EDTA, 4 mM EGTA, 4 mg bovine serum albumin (Fraction V)/ml.
2. [^{3}H]Ins(1,4,5)P_3 is obtained from Amersham International p.l.c., product number TRK 999, specific activity 30–50 Ci/mmol. A portion (100 μl) is diluted with 5400 μl of H_2O: a 25 μl portion should contain no less than 3000 c.p.m. If necessary more [^{3}H]Ins(1,4,5)P_3 is added rather than have insufficient radioactivity.
3. Ins(1,4,5)P_3 standard should be of the highest purity available. A 4 μM and a 250 nM stock are prepared. These can be stored at −20°C for a number of months but freeze-thawing should be avoided.

The samples are thoroughly mixed and then incubated, on ice, for a minimum of 20 min. Incubations are terminated by centrifugation at 12,000 *g* (4°C) for 3 min in, for example, a Hettitch Mikro Rapid/K refrigerated microcentrifuge. Centrifugation time can be increased (e.g., to 5 min) if the pellet is soft and unstable. The supernatant is removed by aspiration and the pellet dissolved in 1 ml of scintillation fluid (Optiphase "Hisafe 3", LKB Wallac).

The Standard Curve

Incubations are conducted in duplicate. Successive 1:1 dilutions of the 250 nM stock Ins(1,4,5)P_3 solution are made to obtain stock solutions of 125, 62.5, 31.25, 15.63, 7.82, 3.91, 1.96 and 0.98 nM Ins(1,4,5)P_3. A portion (25 μl) of the standard solutions is put into labelled tubes; thus, 25 μl of 250 nM stock = 6.25 pmol of Ins(1,4,5)P_3, 25 μl of 125 nM stock = 3.12 pmol of Ins(1,4,5)P_3 etc.

In addition, binding of [^{3}H]Ins(1,4,5)P_3 in the absence of added unlabelled Ins(1,4,5)P_3 (i.e. using 25 μl of H_2O only) is also determined. Non-specific binding is determined in the presence of a 25 μl portion of the 4 μM stock, i.e., 100 pmol of Ins(1,4,5)P_3. A typical standard curve is illustrated in Fig. 2.

Calculation of Results

The standard curve is plotted as $\%B/B_0$ versus pmol of unlabelled Ins(1,4,5)P_3, where B_0 is the maximum specific binding (i.e., c.p.m. bound at 0 pmol of unlabelled Ins(1,4,5)P_3 − NSB), B is the specific binding (i.e., c.p.m. bound at a given concentration of unlabelled Ins(1,4,5)P_3 − NSB) and NSB is non-specific binding. Similar values are calculated for the unknown samples. Thus, values of pmol of Ins(1,4,5)P_3 in each sample can then be determined from the standard curve.

As with any binding assay, readings obtained in the central, i.e., linear,

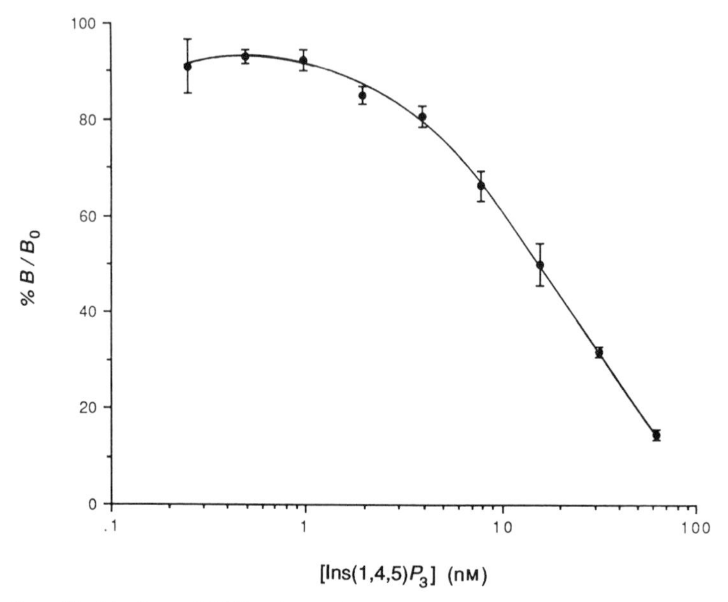

FIG. 2. Specific binding of [^{3}H]Ins(1,4,5)P_3. A typical curve is shown using unlabelled Ins(1,4,5)P_3 as the competing ligand. For %B and B_0, see the text.

part of the standard curve are the most accurate and those that coincide with the extremities of the standard curve are subject to error.

Although not required as part of the assay, it is good practice to calculate routinely %NSB/TC and %B_0/TC, where TC is the total counts per incubation. In general, %NSB/TC should be approximately 10% and %B_0/TC should be between 30% and 50%. If %B_0/TC falls below 30%, it is advisable to use a different frozen portion of binding protein or make a new preparation.

Preparation of Cell Extracts

The quantity of tissue required to obtain values of Ins(1,4,5)P_3 mass that will fall on the linear part of the standard curve varies between cell types. For example, in the authors' experience, a minimum of 10^6 Swiss mouse 3T3 cells are required per sample whereas 2×10^5 NG108-15 neuroblastoma $\times$ glioma cells are sufficient. Cultured cells should be washed and resuspended in a physiological buffer before use, since phenol red (contained in many culture media) can interfere with the binding assay.

Cell extracts can be prepared in a variety of ways. However, it is advisable that recovery measurements are made with radiolabelled Ins(1,4,5)P_3 before a particular method is adopted.

TABLE 1. *Cross-reactivity of [^{3}H]Ins(1,4,5)P_3 binding with inositol and nucleotide phosphates*

Competitor	EC_{50} (M)
Ins(1,4,5)P_3	5.9×10^{-9}
Ins(2,4,5)P_3	1.2×10^{-7}
Ins(1:2-cyclic,4,5)P_3	1.2×10^{-7}
Ins(1,3,4)P_3	No displacement at 5.0×10^{-6}
Ins(1,3,4,5)P_4	1.1×10^{-7}
InsP_5	7.3×10^{-6}
InsP_6	1.3×10^{-5}
ATP	30% displacement at 2.5×10^{-4}
GTP	20% displacement at 2.5×10^{-4}

The Table shows the half-maximal effective concentrations (EC_{50}) of compounds able to compete with 1 nM[^{3}H]Ins(1,4,5)P_3 for its binding site on bovine adrenocortical microsomes. Other inositol phosphates were inactive. Further details can be found in (1) and (12).

Acid extraction yields the highest recoveries of Ins(1,4,5)P_3. Perchloric acid or trichloroacetic acid can be used. Perchloric acid can be removed from the samples by precipitation of potassium perchlorate on addition of KOH in the presence of Hepes and a minimal amount of Universal Indicator (2), whereas trichloroacetic acid is removed by repeated washing with water-saturated diethyl ether (3). Samples prepared using the former method can be added directly to the binding assay whereas samples prepared using trichloroacetic acid may need to be concentrated before assay. In the authors' experience, the use of Freon and tri-*n*-octylamine for neutralization (4) should be avoided since poor recoveries of Ins(1,4,5)P_3 have been reported for this method ((5); and see Chapter 2) and denaturation of the binding protein can occur. However, this method has been used successfully for measurement of Ins(1,4,5)P_3 in brain slices (6). A phenol-based extraction procedure has also been reported to yield good recoveries of Ins(1,4,5)P_3 (7,8) whereas extraction of Ins(1,4,5)P_3 with neutral chloroform/methanol is less efficient (9).

Cell extracts can be prepared either on the day of assay or earlier and stored at −20°C for a number of weeks if necessary. The number of samples that can be comfortably assayed in one day is approximately 100 plus a standard curve.

Generally, 25 μl of cell extract is assayed for Ins(1,4,5)P_3 mass, although smaller volumes can be used and the sample volume adjusted to 25 μl with H_2O. Equally, a larger volume of extract can be assayed, if necessary, by increasing the volume of the other components of the binding assay by equal amounts. Indeed, Challiss et al. (10) suggest that the sensitivity of the binding assay can be increased in this manner. Typical data obtained using the binding assay are illustrated in Fig. 3. As shown in the figure, the assay can be used to determine the intracellular concentration of Ins(1,4,5)P_3.

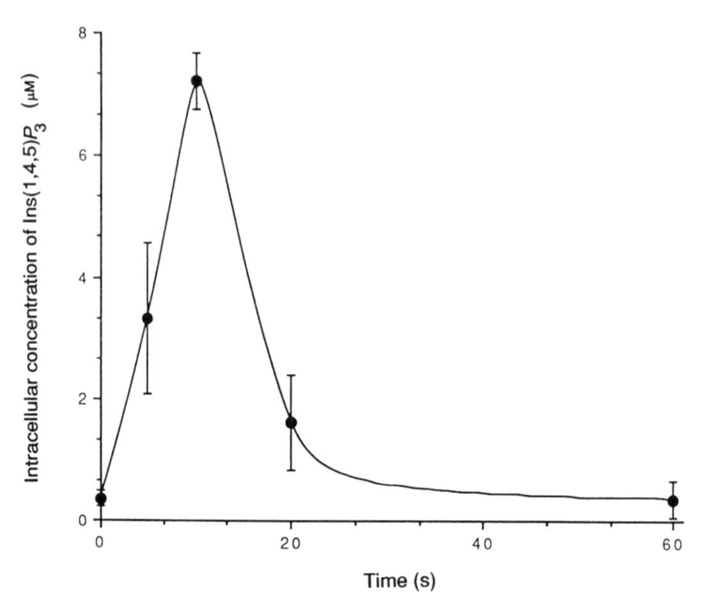

FIG. 3. Time course of Ins(1,4,5)P_3 production in 0.6 μM bombesin-stimulated Swiss 3T3 cells. Incubations were terminated with perchloric acid, centrifuged and the supernatant neutralized with KOH/Hepes (2). A portion of each neutralized sample was assayed for Ins(1,4,5)P_3. Intracellular volume was determined as described in the text. Results are means ± S.D. for data from a single, typical experiment.

The cross-reactivity of other inositol phosphates and nucleotide phosphates (see Table 1) can be assumed to be minimal. Although the cellular concentration of ATP is in the millimolar range, addition of the cell extract to the Ins(1,4,5)P_3 binding assay results in a final concentration of less than 50 μM, which is not sufficient to displace Ins(1,4,5)P_3 from the binding sites.

ESTIMATION OF INTRACELLULAR VOLUME

Intracellular volume can be estimated by preincubation of the cells in the presence of ^3H_2O (0.2 μCi/ml) and [U-^{14}C]sucrose (0.04 μCi/ml) for 10 min (11). Cells in suspension (120 μl) are then centrifuged (12,000 *g* for 20 s) through 500 μl of silicone oil (d = 1.045; a 1:1 mixture of AP100 and AR20, obtained from Wacker Chemicals Ltd, Walton-on-Thames, Surrey, UK) layered above 100 μl of $HClO_4$ (10%, w/v). Radioactivity of the upper and lower phases is determined; [U-^{14}C]sucrose of the lower phase is assumed to represent contamination by the extracellular medium whereas ^3H_2O of the lower phase, after correction for contamination by the extracellular medium, is assumed to be intracellular (87% of which is cytosolic).

The intracellular volume of cells grown on tissue culture dishes can be estimated by preincubation of the cells with the radioisotopes as above, aspiration of the medium and quenching of the cells with 1 M NaOH. A portion of this is then counted and a similar calculation made. Intracellular volume can then be used to calculate the intracellular concentration of Ins(1,4,5)P_3 (see Fig. 3).

TROUBLE SHOOTING

1. Error between replicate samples can be minimized by always using accurate pipettes and using a clean tip for each addition.
2. Once the assay samples have been centrifuged, the supernatant is aspirated as soon as possible. Thus, the pellets are given less time in which to soften and accuracy will be maintained.
3. Scintillant is added and the samples mixed as soon as possible. Pellets, after removal of the supernatant, will harden if allowed to sit on the bench and become impossible to dissolve in the scintillant. Similarly, pellets will not dissolve if allowed to sit in scintillant for any length of time, even a few minutes.
4. Cell extracts prepared with perchloric acid should be carefully neutralized, since high concentrations of salts interfere with the binding assay.

Acknowledgments This work has been supported by grants from the Cancer Research Campaign and the Medical Research Council (UK).

REFERENCES

1. Palmer S, Hughes KT, Lee DY, Wakelam MJO. Development of a novel, Ins(1,4,5)P_3-specific binding assay. Its use to determine the intracellular concentration of Ins(1,4,5)P_3 in unstimulated and vasopressin-stimulated rat hepatocytes. *Cellular Signalling* 1989;1:147–156.
2. Palmer S, Hawkins PT, Michell RH, Kirk CJ. The labelling of polyphosphoinositides with $[^{32}P]P_i$ and the accumulation of inositol phosphates in vasopressin-stimulated hepatocytes. *Biochem J* 1986;238:491–499.
3. Batty IR, Nahorski SR, Irvine RF. Rapid formation of inositol 1,3,4,5-tetrakisphosphate following muscarinic stimulation of rat cerebral cortical slices. *Biochem J* 1985;232:211–215.
4. Sharpes ES, McCarl RL. A high-performance liquid chromatography method to measure ^{32}P-incorporation into phosphorylated metabolites in cultured cells. *Anal Biochem* 1982;124:421–424.
5. Wreggett KA, Irvine RF. A rapid separation method for inositol phosphates and their isomers. *Biochem J* 1987;243:655–660.
6. Challiss RAJ, Batty IR, Nahorski SR. Mass measurements of inositol (1,4,5)trisphosphate in rat cerebral cortical slices using a radioreceptor assay: effect of neurotransmitters and depolarisation. *Biochem Biophys Res Commun* 1988;157:684–691.
7. Hawkins PT, Berrie CP, Morris AJ, Downes CP. Inositol 1,2-cyclic 4,5-trisphosphate is not a product of muscarinic receptor-stimulated phosphatidylinositol 4,5-bisphosphate hydrolysis in rat parotid glands. *Biochem J* 1987;243:211–218.
8. Wong NS, Barker C, Shears SB, Kirk CJ, Michell RH. Inositol 1,2(cyclic),4,5-trisphosphate is not a major product of inositol phospholipid metabolism in vasopressin-stimulated WRK-1 cells. *Biochem J* 1988;252:1–5.

9. Berridge MJ, Dawson RMC, Downes CP, Heslop JP, Irvine RF. Changes in the level of inositol phosphates after agonist-dependent hydrolysis of membrane phospholipids. *Biochem J* 1982;212:473–482.
10. Challiss RAJ, Chilvers ER, Willcocks AL, Nahorski SR. Heterogeneity of [^{3}H]inositol 1,4,5-trisphosphate binding sites in adrenal cortical membranes. *Biochem J* 1990;265:421–427.
11. Shears SB, Kirk CJ. Characterisation of a rapid cellular-fractionation technique for hepatocytes. *Biochem J* 1984;219:375–382.
12. Palmer S, Wakelam MJO. The Ins(1,4,5)P_3 binding site of bovine adrenocortical microsomes: function and regulation. *Biochem J* 1989;260:593–596.

Methods in Inositide Research,
edited by Robin F. Irvine.
Raven Press, Ltd., New York © 1990

11

Separation and Quantification of Isomers of Inositol Phosphates by Ion Chromatography

Grace Y. Sun*, Teng-Nan Lin*, Noel Premkumar**,
Steven Carter†, and Ronald A. MacQuarrie§

**Department of Biochemistry, School of Medicine,
University of Missouri, Columbia, Missouri 65212; **Mobay Chemical Company,
Stanley, Kansas 66085; †Division of Molecular Biology and Biochemistry,
School of Basic Life Sciences, University of Missouri-Kansas City,
Kansas City, Missouri 64110; § and Dionex Corporation,
Sunnyvale, California 94086*

Current interest in understanding the signal transduction mechanism involving turnover of phosphoinositides and generation of diacylglycerols and inositol phosphates has created an urgent need to develop new methods for analysis of these compounds in biological systems. However, the analysis of inositol phosphates has posed special problems because of their low concentrations in biological samples and lack of an identifiable chromophore. As a result, analysis of inositol phosphates has relied mainly on sensitive radiotracer techniques, although the mass of the compounds is not determined by such methods. Since these compounds are metabolically very active, information on specific radioactivity is important for studies on metabolic turnover.

Several procedures have been applied to the separation and quantification of inositol phosphates. The separation of isomers of inositol phosphates is commonly achieved with high-performance liquid chromatography (HPLC) (1), whereas the quantities of inositol monophosphates have been determined by gas chromatography–mass spectrometry (2), and the polyphosphates have been determined by post-column chemical derivatization procedures (3–5). Recently, our laboratory has developed an HPLC procedure together with ion conductivity detection for separation and quantitative measurement of isomeric species of inositol phosphates. This versatile procedure requires no chemical modification reactions, is sensitive in the picomolar range, and can be used to measure the quantities of inositol phosphate isomers and other anions (6). The procedure has been successfully applied to the analysis of inositol phosphates in brain as well as in other tissues and cells.

MATERIALS AND METHODS

Materials

One-month-old male Sprague-Dawley rats (purchased from Taconic Farm, Germantown, NY) are fed regular lab chow and tap water *ad libitum* until the time of the experiment. Inositol monophosphates (Ins*P*), bisphosphates ($InsP_2$) and trisphosphates ($InsP_3$) were obtained from Calbiochem (Richmond, CA) and Sigma Chemical Co. (St Louis, MO). Inositol tetrakisphosphate ($InsP_4$) was from Boehringer Mannheim (Indianapolis, IN). *myo*-[2^3H]Inositol (specific activity, 14 Ci/mmol) was obtained from American Radiolabeled Chemicals (St Louis, MO). Organic solvents and inorganic reagent grade chemicals were from Fisher Scientific (Pittsburgh, PA), and all other biochemicals were from Sigma Chemical Co.

Procedure for Sample Preparation

Approximately one-month-old rats are decapitated and the heads dropped into liquid nitrogen immediately. After immersion in liquid nitrogen for 50 s, the skulls are opened and cerebral cortices are removed and weighed while frozen. The brain tissue (approximately 1 g) is then homogenized at room temperature in 20 ml of $CHCl_3/CH_3OH/12$ M HCl (2:1:0.012, by vol.), extracted with 5 ml of deionized H_2O (HPLC grade), and centrifuged at low speed to separate the phases. The upper aqueous phase is transferred to another test tube and freeze-dried.

Samples prepared by this extraction method are subsequently passed through a maxi-clean IC-H^+ cartridge (Alltech Associates, Deerfield, IL) to reduce the background conductivity. Following this treatment the samples are ready for ion chromatography.

Procedure for Separation and Quantification of Inositol Phosphates by Ion Chromatography

Ion chromatography is performed on a Dionex Bio LC consisting of a Gradient Pump Module (GPM), Liquid Chromatography Module (LCM II), Eluant Degas Module (EDM), and chemically suppressed conductivity detector (CDM II) with an Anion Micro Membrane Suppressor (AMMS) from Dionex Corp. (Sunnyvale, CA). A Dionex Ion Pac AS5A 5μ column and guard column is used for separation of inositol phosphates. The column flow rate is set at 1 ml/min and the Anion Micro Membrane Suppressor is regenerated with 10 ml/min of 0.0125 M sulphuric acid. Peak areas are calculated with a Nelson Analytical 760 intelligent interface and Nelson Analytical Software (Cupertino, CA) adapted to an IBM-AT computer.

Separation of InsP_3

1. The column is washed isocratically with 112.5 mM NaOH: this is accomplished by using 150 mM NaOH solution in reservoir A (eluant A) and deionized H_2O in reservoir B (eluant B) and setting the elution ratio to 75:25. The baseline conductivity at this stage should be approximately 8.5 μS.
2. InsP_3 standard is loaded into the 50 μl sample loop, the conductivity auto-offset and then the standard injected (substantial peak sizes may be observed with a 50–100 pmol sample). The InsP_3 is eluted isocratically with 112.5 mM NaOH. The retention time for inositol 1,3,4-trisphosphate (Ins(1,3,4)P_3) is approximately 13 min and that for Ins(1,4,5)P_3 is approximately 14 min. A standard curve should be constructed with different concentrations of each standard.
3. A sample (50 μl) is applied to the column and eluted in the same way as in step (1). Brain samples normally show Ins(1,4,5)P_3 as the major isomer and only trace amounts of Ins(1,3,4)P_3.
4. The sample is "spiked" with standard to identify the peak better.
5. Peak areas may be obtained from an integrator and concentrations are calculated according to the standard curve constructed in step (2).

Separation of InsP_2

The procedure described above for InsP_3 is used to separate InsP_2 except that the column is pre-equilibrated and eluted isocratically with 67.5 mM NaOH (the ratio eluant A:eluant B is 45:55). The baseline conductivity should be approximately 5 μS, and the retention time for Ins(1,4)P_2 is approximately 8.5 min. Analysis of brain samples for InsP_2 normally shows only one peak, which corresponds to Ins(1,4)P_2.

Separation of InsP

The procedure used to separate Ins$P$ is the same as described above except that 10.5 mM NaOH (the ratio eluant A:eluant B is 7:93) is used to pre-equilibrate the column and to separate the Ins(1)P and Ins(4)P. The conductivity baseline should be approximately 2.5 μS. The retention time for

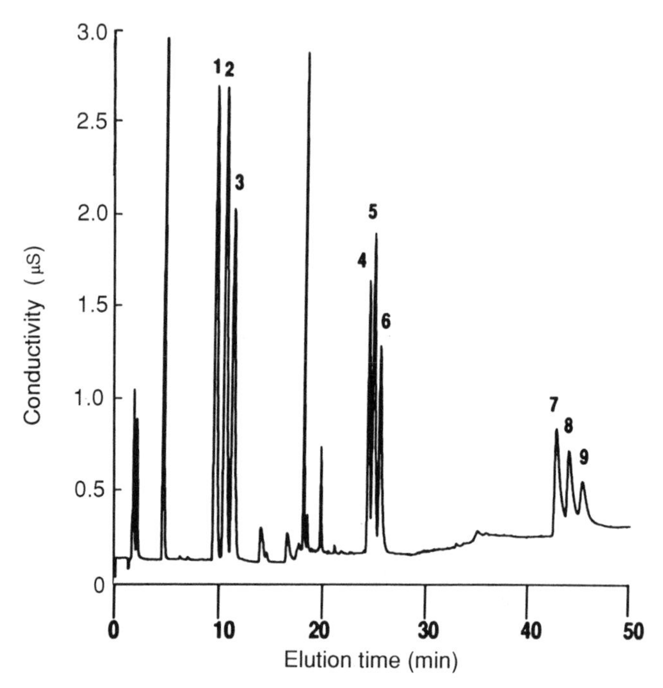

FIG. 1. Separation of isomers of inositol mono-, bis-, and trisphosphate standards by a step gradient elution procedure. The conditions for stepwise gradient elution are: Ins$P$ region (0–18 min), 16.5 mM NaOH; InsP_2 region (18–36 min), 78.0 mM NaOH; and InsP_3 region (36–50 min), 112.5 mM NaOH. The peaks in the figure were obtained following injection of 1.5 μg of each inositol monophosphate isomer and 0.5 μg of each of the other compounds. The peak assignments are as follows: peak 1, Ins(2)P; peak 2, Ins(1)P; peak 3, Ins(4)P; peak 4, Ins(1,4)P_2; peak 5, Ins(2,4)P_2; peak 6, Ins(4,5)P_2; peak 7, Ins(1,3,4)P_3; peak 8, Ins(1,4,5)P_3; and peak 9, Ins(2,4,5)P_3.

Ins(1)P is approximately 13 min and that for Ins(4)P is approximately 14 min. Under normal conditions, brain samples show the presence of both Ins(1)P and Ins(4)P peaks. One of the limitations of this elution scheme is the presence of a large unknown peak which elutes just prior to the Ins(1)P peak. In some instances, these two peaks are not well separated (see Fig. 3).

Construction of Standard Curves

Different amounts of standards are applied to the column and eluted with the appropriate solvents, as described above. The peak area is then plotted against the amount (pmol) of the sample.

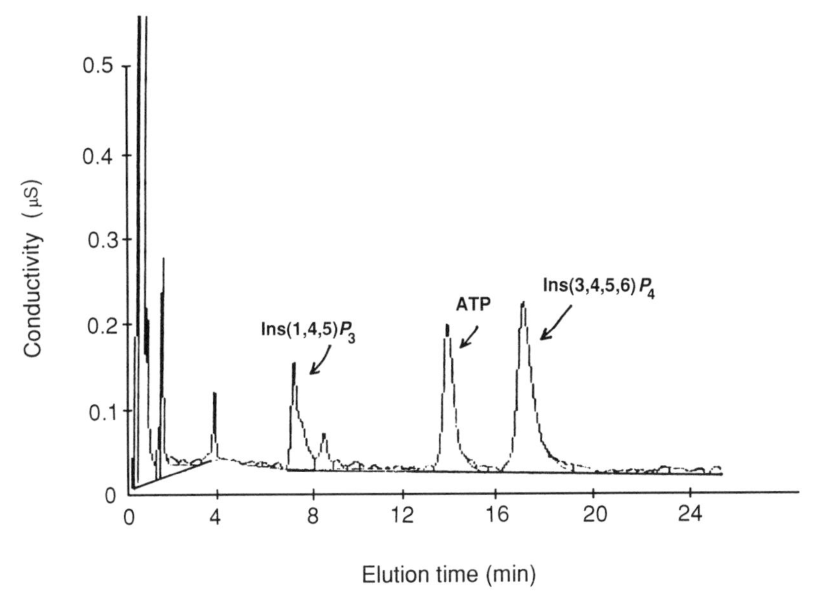

FIG. 2. Separation of Ins(1,4,5)P_3, ATP, and Ins(3,4,5,6)P_4. A mixture consisting of 150 pmol of each compound was separated by ion chromatography using the conditions described in Materials and Methods except that the NaOH concentration was 123 mM.

NOTES ON THE SEPARATION PROCEDURE

Sample Preparation

1. Repetitive freezing and thawing of the samples appears to decrease the concentrations of inositol phosphates.
2. Freeze-drying is the preferred method for reduction of sample size after extraction of the tissues. Alternatively, a rotary evaporator may be used for removal of organic solvent and the residue suspended in water and centrifuged to remove insoluble material.
3. However, other extraction procedures, including those which use neutral chloroform–methanol combinations give analytical values which are similar to those reported below. An extraction system using hexane/acetone (3:2, v/v) gave consistently low values for InsP_3.

Elution and Identification

1. A step gradient elution program may be used to elute sequentially Ins$P$, InsP_2 and InsP_3 (Fig. 1), although baseline subtraction may be necessary

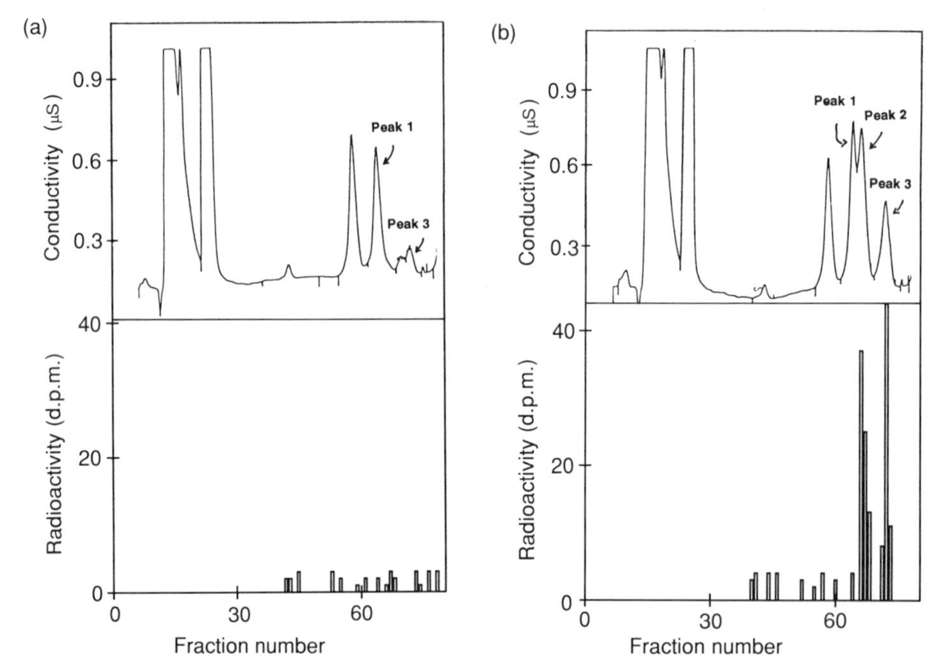

FIG. 3. Analysis of brain samples for Ins(1)*P* and Ins(4)*P*. Brain samples were obtained from control rats or from rats treated with lithium (7 mequiv./kg body wt 4 h prior to sacrifice). Rats were injected 16 h prior to sacrifice with [^{3}H]inositol to prelabel inositol phopsholipids. **(a)** Chromatograms of the Ins*P* region obtained from control animals. The lower portion of the figure shows the radioactivity in d.p.m. of [^{3}H]inositol phosphates corresponding to the elution profile. **(b)** Chromatograms obtained from animals treated with lithium. The chromtograms show that both Ins(1)*P* (peak 2) and Ins(4)*P* (peak 3) increase in size in the presence of lithium. The lower portion of the figure shows that these peaks correspond to the presence of radioactivity in the form of [^{3}H]inositol phosphates.

to provide best results. Alternatively, isocratic elution procedures may be employed for individual classes of inositol phosphates. In this latter case, it is usually more convenient to elute the samples in the order InsP_3, InsP_2 and Ins$P$, because it is easier to equilibrate the column with high NaOH concentrations first.

2. It is possible to separate InsP_4 and other polyanions by eluting with a higher concentration of NaOH (123 mM). Figure 2 shows the elution profile for separation of a mixture consisting of 150 pmol each of Ins(1,4,5)P_3, ATP, and Ins(3,4,5,6)P_4. The approximate elution times for the compounds are: Ins(1,4,5)P_3, 7.6 min; ATP, 13.5 min; and Ins(3,4,5,6)P_4, 17 min. There are some indications that repeated elution of these phosphates (with high NaOH concentrations) may dramatically decrease column efficiency.

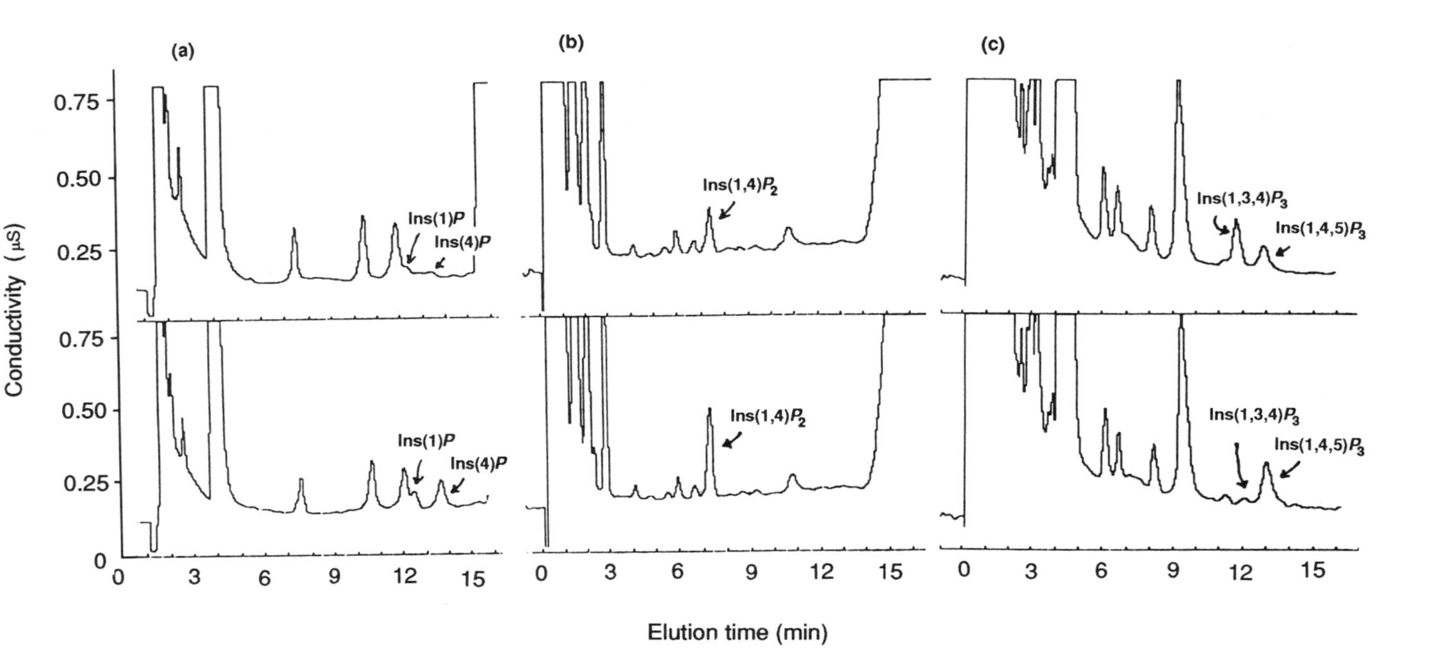

FIG. 4. Analysis of brain samples for inositol mono-, bis- and trisphosphates. **(a)** Chromatograms depicting Ins(1)P and Ins(4)P in brain samples (upper) and in brain samples after spiking with 22.5 pmol of Ins(1)P and 50 pmol of Ins(4)P standard (lower). Isocratic elution was used as described in Materials and Methods. **(b)** Chromatograms depicting Ins(1,4)P_2 peak in brain (upper) and after spiking with 73.5 pmol of Ins(1,4)P_2 standard (lower). Ins(1,4)P_2 concentrations were determined to be 10 ($\pm$ 3) nmol/g wet wt (N = 4) in control samples. **(c)** Chromatograms depicting Ins(1,3,4)P_3 and Ins(1,4,5)P_3 in brain. The upper chromatogram is spiked with 114 pmol of Ins(1,3,4)P_3 and the lower graph is spiked with 57 pmol of Ins(1,4,5)P_3. The concentration of Ins(1,4,5)P_3 in control samples was found to be 5.8 ($\pm$ 0.8) nmol/g wet wt (N = 10).

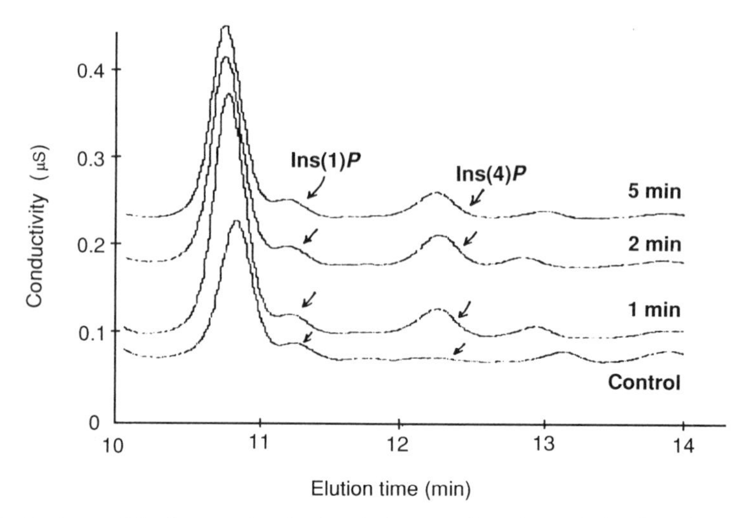

FIG. 5. Analysis of brain samples for inositol monophosphates following decapitation-induced ischemia. Following decapitation, rat brains were incubated at 37°C for the indicated times, then frozen in liquid nitrogen, and extracted as described in Materials and Methods. The chromatograms show that decapitation-induced ischemia results in a substantial increase in the amounts of Ins(4)*P*, but little change in the amounts of Ins(1)*P*.

3. Samples should be spiked with standards periodically to check for consistency of retention times.
4. After use, it is necessary to wash the column with 0.5 M NaOH with a flow rate of 0.4 ml/min.
5. The chromatographic columns deteriorate with time after extensive use as evidenced by the inability to separate Ins(1)*P* from contaminating material in brain samples (peaks 1 and 2 of Fig. 3).

Analysis of Biological Samples

1. The limits of detection are approximately 10 pmol for inositol phosphates when using the 50 μl injection loop of the liquid chromatograph.
2. Brain tissue homogenate may be spiked with known amounts of inositol phosphate derivatives to determine the extent of recovery after extraction and ion chromatography. Our percentage recoveries for Ins(4)*P*, Ins(1,4)P_2, and Ins(1,4,5)P_3 were 80.1, 87.2 and 76.9, respectively.
3. The concentrations of Ins(1)*P* and Ins(4)*P* in control samples from brain are 8.2 (±0.6) and 4.0 (±0.3) nmol/g wet wt ($N = 4$), respectively. In lithium-treated animals, the concentrations of Ins(1)*P* and Ins(4)*P* are increased over control values by 40-fold and 20-fold, respectively (Fig. 3).
4. Quantification of Ins(1)*P* is the most difficult to achieve of the common

inositol phosphate isomers because of the presence of a large unknown peak which elutes just prior to that of Ins(1)P (Fig. 3). Inositol bis- and trisphosphates are eluted under conditions in which there is little interference from contaminating ions (Fig. 4).

5. Post-decapitative ischemia results in substantial increases in Ins(4)P, as shown in Fig. 5. After 2 min of ischemia, Ins(4)P concentrations increase approximately five-fold over control values, whereas Ins(1)P concentrations change little. This ischemic treatment results in a transient appearance of Ins(1,4,5)P_3 within the first 30 s and approximately six-fold increases in Ins(1,4)P_2 after 2 min (data not shown).

Acknowledgment Supported in part by grant AA-06661 from the NIAAA.

REFERENCES

1. Irvine RF, Anggard EE, Letcher AJ, Downes CP. Metabolism of inositol 1,4,5-trisphosphate and inositol 1,3,4-trisphosphate in rat parotid glands. *Biochem J* 1985;229:505–511.
2. Honchar MP, Olney JW, Sherman WR. Systemic cholinergic agents induce seizures and brain damage in lithium treated rats. *Science* 1983;202:323–325.
3. Meek JL, Nicoletti F. Detection of inositol trisphosphate and other organic phosphates by high performance liquid chromatography using an enzyme-loaded post column reactor. *J Chromatogr* 1986;351:303–311.
4. Minear RA, Segers JE, Elwood JW, Mulholland PJ. Separation of inositol phosphates by high performance ion exchange chromatography. *Analyst* 1988;113:645–649.
5. Pittet D, Schlegel W, Lew PD, Monod A, Mayr GW. Mass changes in inositol tetrakis- and pentakisphosphate isomers induced by chemotactic peptide stimulation in HL 60 cells. *J Biol Chem* 1989;264:18,489–18,493.
6. Smith RE, MacQuarrie RA. Determination of inositol phosphates and other biologically important anions by ion chromatography. *Anal Biochem* 1988;170:308–315.

Methods in Inositide Research,
edited by Robin F. Irvine.
Raven Press, Ltd., New York © 1990

12

Purification of Phosphatidylinositol 4-Phosphate and Phosphatidylinositol 4,5-Bisphosphate by Column Chromatography

Martin G. Low

Rover Physiology Research Laboratories, Department of Physiology and Cellular Biophysics, College of Physicians and Surgeons of Columbia University, New York, New York 10032

This chapter describes column chromatographic methods for the purification of the polyphosphoinositides phosphatidylinositol 4-phosphate and phosphatidylinositol 4,5-bisphosphate. These methods were developed from an existing procedure (1) in order to provide polyphosphoinositide substrates for phospholipase C assays (2,3). The purification of milligram amounts of the polyphosphoinositides by precipitation with methanol and acid treatment of a bovine brain lipid extract followed by chromatography on either a DEAE-cellulose column or a preparative high-performance liquid chromatography (HPLC) amino column is described in the next section. The purification of radioactively labelled polyphosphoinositides on a smaller HPLC amino column is described later.

PURIFICATION OF PHOSPHATIDYLINOSITOL 4-PHOSPHATE AND PHOSPHATIDYLINOSITOL 4,5-BISPHOSPHATE FROM BOVINE BRAIN

Extraction of Lipids

The author's use of this procedure involved exclusively a lipid extract (Folch fraction I) of bovine brain that was obtained from a commercial source (Sigma Chemical Company, St Louis, MO) but the method should be applicable to a freshly prepared lipid extract from brain or other tissues. Bovine brain Folch fraction I (1 g) is dissolved in 12 ml of chloroform. The phosphoinositides are precipitated by mixing with 22 ml of methanol followed by centrifugation at 1000 *g* for 5 min. The supernatant is discarded, and the pellet redissolved in 12 ml of chloroform. This precipitation with methanol is done a total of six times.

The final "methanol precipitate" is dissolved in 15 ml of chloroform, 15 ml of methanol, 5 ml of 1 M HCl. Additional 1 M HCl (8.5 ml) is added to give two phases. After centrifugation the upper phase is discarded and, to the lower phase, 15 ml of methanol and 5 ml of 2 M NaCl are added and the sample mixed until it becomes clear. The phases are separated by the addition of 2.5 ml of 2 M NaCl and 6 ml of H_2O. The upper phase is removed and the lower phase washed once more with methanol and 2 M NaCl, as above (the above steps are all done at 0–4°C). The lower phase is transferred to a clean tube, evaporated to dryness under nitrogen gas, the lipids redissolved in 2 ml of chloroform and stored at −20°C until required. If the chromatographic step is to be performed within a few days, a convenient volume of solvent A (chloroform/methanol/H_2O, 20:9:1, by vol.) may be used to dissolve the lipids.

This procedure may be interrupted after the precipitation with methanol (the precipitate is dissolved in chloroform and stored at −20°C) but the acid and NaCl washes should be completed as rapidly as possible to minimize acid hydrolysis. One g of starting material yields 150–250 μmol of precipitated phospholipid (as determined by organic phosphorus).

Chromatography on DEAE-Cellulose

DEAE-cellulose (100 g) (Sigma Chemical Company, St Louis, MO) is washed on a filter with 500 ml of 1 M NaOH, H_2O to neutrality, 500 ml of 10% (v/v) acetic acid and finally H_2O to neutrality. The majority of the H_2O is removed by stirring the DEAE-cellulose for about 5 min with two volumes of methanol and then allowing the mixture to stand for approximately 30 min. The methanol is decanted and the treatment repeated twice with methanol and then twice with two volumes of solvent A. The DEAE-cellulose is finally resuspended in solvent A and packed into a glass column with solvent-resistant fittings (2 cm × 38 cm). It is not necessary to repeat all of this washing procedure after each use but the column should be eluted with at least 350 ml of solvent A prior to each run. Because of the relatively high density (relative to DEAE-cellulose) and vapour pressure of solvent A, the DEAE-cellulose bed is mechanically unstable and is prone to trap large vapour bubbles within the bed. This is mainly a problem in the absence of solvent flow (i.e., at the end of a run) or if the ambient temperature is high. These problems can be minimized by carrying out the chromatography in a cold room and/or repacking the column just before use.

The brain phospholipids (prepared as described above) are dissolved in

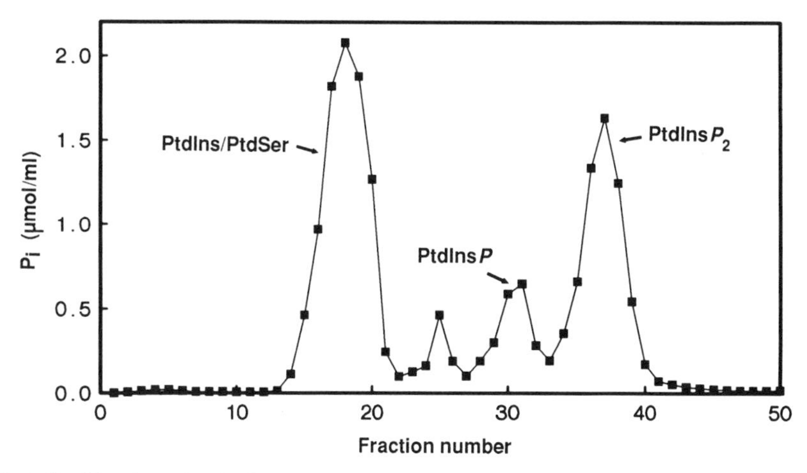

FIG. 1. Purification of polyphosphoinositides from bovine brain on a DEAE-cellulose column. Acid-washed methanol-precipitated Folch Fraction I was eluted from a DEAE-cellulose column as described in the text. The peak eluting at fraction no. 25 was reproducibly present but was not identified. The composition of the other peaks was determined by thin-layer chromatography analysis of the pooled fractions. PtdIns, phosphatidylinositol; PtdSer, phosphatidylserine; PtdIns*P*, phosphatidylinositol 4-phosphate; PtdInsP_2, phosphatidylinositol 4,5-bisphosphate.

50 ml of solvent A and applied to the DEAE-cellulose column. The column is then eluted at a flow rate of approximately 100 ml/h with a 1 liter linear gradient from 100% solvent A to 100% solvent B (solvent A containing 0.6 M ammonium acetate) and 13–15-ml fractions collected. The fractions are assayed for organic phosphorus and the peak fractions (usually four from each peak) are pooled and filtered. Average recovery of applied organic phosphorus is approximately 90%. A typical elution profile is shown in Fig. 1.

Chromatography on an HPLC Amino Column

Alternatively, the chromatographic step can be performed by preparative HPLC. The brain phospholipids (prepared as described above) dissolved in 5 ml of solvent A are applied at a flow rate of 2.5 ml/min to a 10 mm × 250 mm amino-NP column (5 μm spherical silica with *n*-propylamine bonded phase; IBM Instruments, Inc., Danbury, CT) with a 4.5 mm × 50 mm guard column (same packing material) equilibrated with solvent A. The column is eluted at a flow rate of 2.5 ml/min with 25 ml of solvent A, a 125 ml gradient of 100% solvent A to 100% solvent B, and 100 ml of 100% solvent B; 5-ml fractions are collected. The fractions are assayed for organic phosphorus and peak fractions (usually three from each peak) are pooled. Average re-

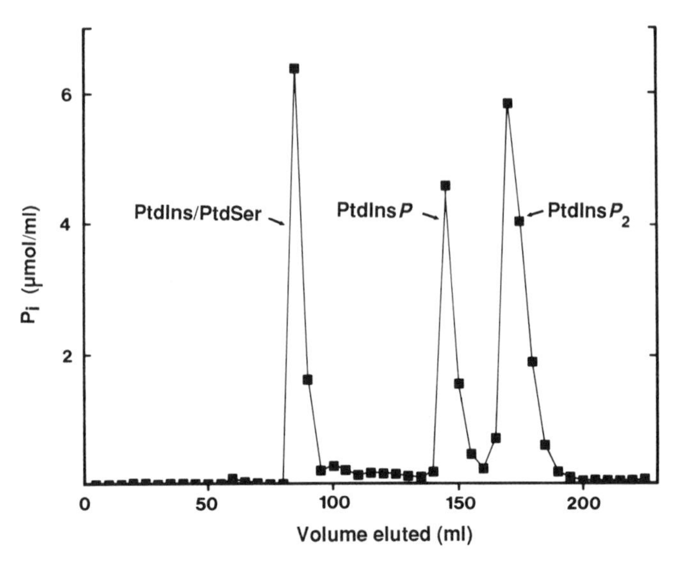

FIG. 2. Purification of polyphosphoinositides from bovine brain on an HPLC amino column. Acid-washed methanol-precipitated Folch fraction I was eluted from an HPLC amino column as described in the text. Composition of the PtdIns*P* and PtdInsP_2 peaks were determined by thin-layer chromatography analysis of pooled fractions. Composition of the Ptd/Ins/PtdSer peak was inferred from a comparison with Fig. 1. For abbreviations, see legend to Fig. 1.

covery of applied organic phosphorus is approximately 80%. A typical elution profile is shown in Fig. 2.

Concentration of Pooled Fractions

The volume of the pooled fractions is adjusted to 18 ml with solvent A (HPLC method) or to 54 ml, which is then split into three 18 ml portions (DEAE-cellulose method). To each 18 ml portion (in a 50 ml glass tube), methanol (6.6 ml) and 2 M NaCl (10.2 ml) are added to separate the phases. The lower phase is washed with 12 ml of methanol and 10.8 ml of 2 M NaCl. The above steps are done at 0–4°C. The lower phase is removed, evaporated to dryness under nitrogen gas in a clean tube and dissolved in 5 ml of chloroform, 0.7 ml of methanol, 0.05 ml of H_2O. The purified polyphosphoinositides (in contrast to the crude starting material) do not dissolve readily in chloroform or chloroform:methanol mixtures. Addition of a small amount of water is therefore essential to dissolve the purified lipids at the 2–3 mM level. The composition of the final solvent given here was chosen to minimize the amount of methanol and H_2O so that solvent can be rapidly removed by evaporation for preparation of substrates. The purified polyphosphoinositides will also dissolve in solvent A at this concentration. These procedures produce approximately 7–15 μmol of phosphatidylinositol 4-phosphate and 10–20 μmol of phosphatidylinositol 4,5-bisphosphate.

PURIFICATION OF ^{32}P-LABELLED PHOSPHATIDYLINOSITOL 4-PHOSPHATE AND PHOSPHATIDYLINOSITOL 4,5-BISPHOSPHATE FROM HUMAN PLATELETS

The procedure described here utilizes human platelets as a convenient source of ^{32}P-labelled phospholipids but the procedure should in principle be adaptable to phospholipid from other tissues and cell types. Platelets prepared from 60 ml of fresh human blood are incubated in 3 ml of 140 mM NaCl, 5 mM KCl, 0.05 mM $CaCl_2$, 0.1 mM $MgCl_2$, 16.5 mM glucose, 0.1 mg bovine serum albumin/ml, 15 mM Hepes (pH 7.4), containing 2–6 mCi of [^{32}P]P_i (carrier-free, ICN Radiochemicals, Irvine, CA) for 2 h at 37°C with occasional gentle shaking. Prostaglandin E_1 (PGE_1) (2.8 μM) is also added, since it prevents activation and increases the yield of [^{32}P]-polyphosphoinositide. The lipids are extracted with 11.25 ml of chloroform/methanol/HCl (50:100:1.3, by vol.) for 30 min at 25°C. Chloroform (3.75 ml) and 0.1 M HCl (3.75 ml) are then added to separate the phases. The lower phase is washed twice with 5 ml of methanol, 4.5 ml of 2 M NaCl, 0.5 ml of 100 mM EDTA-NaOH (pH 7.4) and then evaporated to dryness in a clean tube with approximately 1.5 μmol of acid-washed, methanol-precipitated bovine brain phosphoinositides (material prepared as above) to act as a carrier. The extract is finally redissolved in 0.5 ml of solvent A and applied at a flow rate of 1 ml/min to a 4.5 mm × 250 mm amino-NP column (same packing as column mentioned above, p. 147) equilibrated in solvent A. The column is eluted with 5 ml of solvent A, a 25 ml linear gradient from 100% solvent A to 100% solvent B, and 20 ml of solvent B. One-ml fractions are collected and radioactivity determined by liquid scintillation counting. Average recovery of applied radioactivity is approximately 70% (a typical elution profile is shown in Fig. 3).

Peak fractions are pooled, the volume adjusted to 6 ml with solvent A and methanol (2.2 ml) and 3.4 ml of 2 M NaCl added to separate the phases. The lower phase is washed with 4 ml of methanol and 3.6 ml of 2 M NaCl. The above steps are done at 0–4°C. The lower phase is evaporated to dryness in a clean tube and finally redissolved in 5 ml of chloroform, 0.7 ml of methanol, 0.05 ml of H_2O. Approximately 2–6 μCi of [^{32}P]phosphatidylinositol 4-phosphate and 1–3 μCi of [^{32}P]phosphatidylinositol 4,5-bisphosphate are obtained by this procedure.

COMMENTS ON THE PURIFICATION PROCEDURE

The DEAE-cellulose chromatography procedure described here is a modification of an earlier method devised by Hendrickson and Ballou (1). The

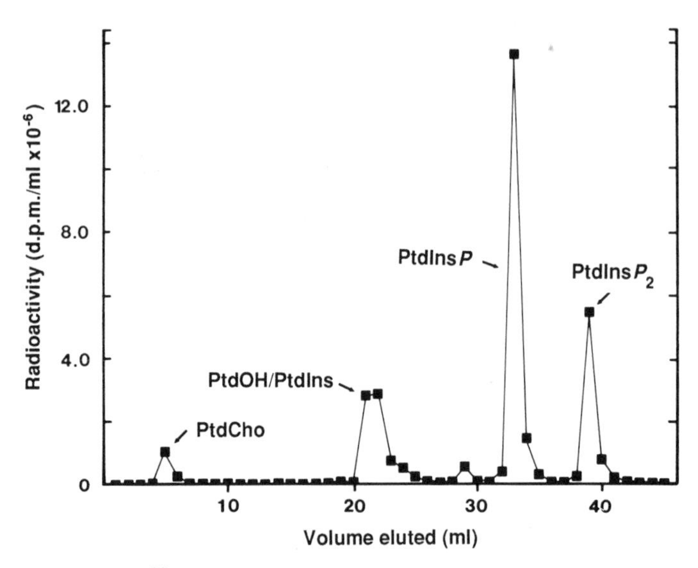

FIG. 3. Purification of ^{32}P-labelled polyphosphoinositides from human platelets on an HPLC amino column. The lipid extract of ^{32}P-labelled human platelets was eluted from an HPLC amino column as described in the text. Composition of the Ptd*Ins*P and PtdInsP_2 peaks were determined by thin-layer chromatography analysis and autoradiography of pooled fractions. Composition of the other two peaks was inferred from the mobility of standard lipids. PtdCho, phosphatidylcholine; PtdOH, phosphatidic acid; d.p.m., disintegrations per minute. For other abbreviations, see the legend to Fig. 1.

major difference between the two methods is the pretreatment of the material loaded on to the column. In the procedure of Hendrickson and Ballou (1) the lipids were treated (in aqueous suspension) with the sodium salt of EDTA followed by chromatography on a Chelex 100 column. However, a substantial proportion of the phosphatidylinositol 4,5-bisphosphate was not retained by the DEAE-cellulose column; an observation that these authors attributed to the persistence of a mixed Ca^{2+}/Mg^{2+} chelate of this phosphoinositide in which the electrical charge of the lipid was neutralized (1). Since none of this chelate is obtained using the method described here (note absence of P_i-containing material in fractions 5–10, Fig. 1) it is likely that the chelate is destroyed by the acid treatment of the methanol-precipitated Folch fraction before it is applied to the column. It appears, therefore, that an intermediate treatment with acid is a more effective method for conversion of this chelate to the sodium salt.

The same eluting solvents are used in the HPLC methods; the only major difference is that an HPLC amino column was substituted for the DEAE-cellulose ion exchange column. Although amino columns are relatively weak anion exchangers, the separation obtained here (compare Figs. 2 and 3 with Fig. 1) also appears to be on the basis of ionic charge. The inositol phospholipids are eluted in order of increasing degree of phosphorylation, and neutral phospholipids such as phosphatidylcholine appear to pass straight through the column. However, it is of interest to note that phosphatidylinositol 4,5-bisphosphate elutes during the gradient on the DEAE-cellulose column but requires isocratic elution with 100% of solvent B to elute it from the amino columns. This suggests that other interactions may be responsible for retention by the amino columns in addition to ionic ones. We have observed no deterioration in resolution with repeated use of the HPLC columns; i.e., approximately 20 and 5 times for the 4.5 mm × 250 mm and 10 mm × 250 mm columns, respectively.

The HPLC procedures described here are clearly not suitable for preparation of [^{32}P]phosphatidylinositol since phosphatidic acid, which in most cell types is more rapidly labelled by [^{32}P]P_i, is not resolved from it (see Fig. 3). Furthermore, it is likely that other anionic lipids such as phosphatidylserine will be poorly resolved from phosphatidylinositol by the amino column, as observed with the DEAE-cellulose procedure (see Fig. 1 and (1)).

For most purposes the HPLC technique will be superior in resolution, speed and overall convenience to the DEAE-cellulose procedure and is the method of choice if HPLC equipment is available. DEAE-cellulose is relatively cheap and larger-scale purifications might require the use of this matrix. However, the capacities of the two columns were not determined, so their relative cost-effectivenesses cannot be compared. The HPLC procedure may also be superior in some or all of these aspects to alternative procedures for purification of polyphosphoinositides, e.g., preparative thin-layer chromatography (see Chapter 13) or neomycin affinity chromatography (see Chapter 1). However, since we do not have extensive direct experience with these other preparative techniques, we cannot comment on this possibility. The reader should be aware that relatively little effort was devoted by the author to refining and optimizing the HPLC technique and it is probable that minor modifications to the procedure will improve significantly some aspect of performance.

Acknowledgments The author's work is currently supported by grants from the National Institutes of Health, the American Heart Association (Wyeth-Ayerst awardee) and Samuel W. Rover.

REFERENCES

1. Hendrickson HS, Ballou CE. Ion exchange chromatography of intact brain phosphoinositides on diethylaminoethyl cellulose by gradient salt elution in a mixed solvent system. *J Biol Chem* 1964;239:1369–1373.
2. Low MG, Weglicki WB. Resolution of myocardial phospholipase C into several forms with distinct properties. *Biochem J* 1983;215:325–334.
3. Low MG, Carroll RC, Cox AC. Characterization of multiple forms of phosphoinositide-specific phospholipase C purified from human platelets. *Biochem J* 1986;237:139–145.

Methods in Inositide Research,
edited by Robin F. Irvine.
Raven Press, Ltd., New York © 1990

13

Separation of Phosphoinositides and Other Phospholipids by High-Performance Thin-Layer Chromatography

Grace Y. Sun and Teng-Nan Lin

Department of Biochemistry, School of Medicine, University of Missouri, Columbia, Missouri 65212

Recent interest in elucidating the receptor-mediated signal transduction mechanism involving hydrolysis of polyphosphoinositides has necessitated the development of more sophisticated methods for the separation and analysis of individual phophoinositides (phosphatidylinositol (PtdIns), phosphatidylinositol phosphate (PtdIns*P*) and phosphatidylinositol bisphosphate (PtdInsP_2)). In some instances, these agonists not only stimulate the hydrolysis of PtdInsP_2 but may also give rise to an increased turnover of other phospholipids. For example, an increase in the synthesis of phosphatidic acid (PtdOH) is shown frequently during agonist stimulation of PtdInsP_2 hydrolysis. There is also evidence for agonist stimulation of phosphatidylcholine hydrolysis either through the action of phospholipase A_2 or phospholipase D. In fact, the increase in phospholipase D activity may account for some of the PtdOH formed. Therefore, it is important to be able to examine the phosphoinositides together with other phospholipids during agonist stimulation.

For a number of years, researchers in our laboratory have used two-dimensional high-performance thin-layer chromatography (HPTLC), with intermediate acid reaction, for the separation of membrane phospholipids and plasmalogens (1). Although this procedure can separate PtdIns from phosphatidylserine (PtdSer) and PtdOH, the polyphosphoinisitides are not separated and remain at the origin. Subsequently, we have attempted to modify this procedure to include separation of PtdIns*P* and PtdInsP_2 (2). Because polyphosphoinositides and PtdOH are normally present in trace amounts in tissues and cells, this procedure is not suitable for mass determination but can be applied successfully to studies in which tissues, membranes or cells are prelabelled with $[^{32}P]P_i$, $[^{3}H]$inositol or $[^{14}C]$arachidonic acid. In this chapter, the procedures for labelling brain phospholipids as well as those for extraction and separation of the labelled phospholipids by the improved HPTLC technique are described.

MATERIALS AND METHODS

Labelling of Brain Phospholipids with [^{32}P]P$_i$ or [^{32}P]ATP

The phospholipids in rat or mouse brain can be labelled by injecting intracerebrally with 50 μCi of [^{32}P]P$_i$ or 10 μCi of [^{32}P]ATP (New England Nuclear, Boston, MA). Injections are made using a 27 gauge × ⅛ in. needle adapted with tubing to give a 3 mm penetration. Animals are killed by decapitation at specific times after injection and the heads are frozen immediately by dropping them into liquid nitrogen. In studies involving isolation of brain subcellular membranes, the brain cortex is removed immediately after decapitation and dropped into ice-cold sucrose solution. In order to be consistent in this procedure with different samples, a 30 s period is allowed between decapitation and dissection of brain tissue.

Extraction of Brain Lipids

Brain tissue is suspended in 10 volumes of ice-cold 0.32 M sucrose containing 50 mM Tris·HCl (pH 7.4) and 1 mM EDTA and homogenized with a motor-driven glass homogenizer equipped with a Teflon pestle. Subcellular fractionation is carried out according to the procedure described by Sun et al. (3). Portions of the brain homogenate or isolated membranes are subjected to a two-step lipid extraction procedure. First, the tissue homogenate is extracted with four volumes of chloroform/methanol (2:1, v/v). (Note: the volumes of the organic and aqueous layers may have to be proportionally adjusted depending on the amount of tissue.) After vortex and brief centrifugation, the lower organic phase is carefully removed and transferred to another test tube. The aqueous phase is further extracted with two volumes of chloroform/methanol/12 M HCl (2:1:0.012, by vol.). The second acidic extraction procedure is important for a more complete recovery of the acidic phospholipids, especially the polyphosphoinositides. The organic layer obtained from the acidic chloroform/methanol extract is subsequently neutralized with one drop of 4 M NH_4OH before being combined with the first organic extract. This neutralization step is needed to protect the plasmalogens, which are sensitive to strong acids. The combined organic solvent is evaporated to dryness in a rotary evaporator. The lipids are resuspended in chloroform/methanol (2:1, v/v) and stored at −20°C until use.

Separation of Phospholipids by HPTLC

Brain lipid extract equivalent to 1–1.5 mg of protein is most suited to this type of HPTLC separation. The silica gel 60 HPTLC plates (10 cm × 10 cm, E. Merck, Darmstadt, FRG) should be impregnated with oxalate (by dipping) and then dried overnight before use. The oxalate solution consists of 1% (w/v) potassium oxalate, 2 mM Na_2EDTA; it is mixed with methanol in the ratio oxalate/methanol 2:3 (v/v). In most instances, it is necessary to add polyphosphoinositide standards (100 μg) to the samples for visualization purposes. (The polyphosphoinositide mixture from Sigma Chemical Co., St Louis, MO, may be used.)

Procedure for Development of the HPTLC Plates

1. The lipid sample is applied to the lower left corner of the HPTLC plate using a disposable pipette. After sample application, the HPTLC plate is dried briefly by blowing with, for example, a hair drier (see note below about temperature settings) to remove residual moisture from the plate.
2. The first solvent system comprises chloroform/methanol/acetone/16 M NH_4OH (70:45:10:10, by vol.). The solvent is allowed to move for 9 cm up the plates. After development, it is necessary to dry the plates with a hair drier for 5 min. The first solvent system will result in migration of most phospholipids except polyphosphoinositides.
3. The HPTLC plates are subsequently developed in the same direction with a second solvent system consisting of chloroform/methanol/16 M NH_4OH/H_2O (36:28:2:6, by vol.). The solvent is allowed to move up about 7 cm. The second solvent system is used to separate the PtdIns*P* and PtdInsP_2 from the origin. After development with the second solvent system, the HPTLC plates are thoroughly dried by blowing with a hair drier for 6–8 min.
4. The plates are exposed to HCl fumes for 3 min. In our laboratory, we use a rectangular glass tray with lid so that the HPTLC plate (10 cm × 10 cm) can be placed face down on top of the tray. The tray is equilibrated with 20 ml of fuming HCl, with the lid on, for 10 min before use. (Caution: the entire procedure should be carried out in a good fume cupboard.) Immediately after HCl exposure, excess HCl fumes should be removed from the plates by blowing with a hair drier for 8 min.
5. The plates are inverted through 90° and placed in a third solvent system consisting of chloroform/methanol/acetone/acetic acid/0.1 M ammonium acetate (140:60:55:4.5:10, by vol.). After development, the plates are dried briefly with a hair drier.

Visualization of Lipid Spots

Visualization of lipid spots may be carried out either by exposing the plate to iodine vapours or by spraying the plate with 2′,7′-dichlorofluorescein (1% in methanol) and viewing under an ultraviolet lamp. The iodine procedure is most suitable for identifying the lipid spots for measurement of radioactivity, whereas the fluorescent dye procedure is used for recovering the phospholipids for analysis of the acyl groups by gas–liquid chromatography. After exposure to iodine vapours, the lipid spots are scraped into counting vials for measurement of radioactivity using a Beckman LS5800 scintillation spectrometer. For a permanent record of the lipid spots, the plates may be sprayed with a solution containing 3% (v/v) copper acetate in 8% (v/v) aqueous phosphoric acid and subsequently heated for 20 min at 140°C (charring).

Autoradiography

For autoradiography, an unstained plate may be exposed to Kodak OMAT-AR film at −70°C overnight (exposure time can be shortened or lengthened depending on the amount of radioactivity on the plate). After developing the film, the plate can be stained and lipid spots scraped off for counting of radioactivity.

Examples of HPTLC Separation of Brain Phospholipids Prelabelled with [^{32}P]ATP

Examples of the HPTLC separation of the phospholipids in brain homogenate and myelin are shown in Fig. 1. Since these samples were prelabelled with [^{32}P]ATP, autoradiograms of the corresponding samples are shown on the right. Due to the small amounts of polyphosphoinositides present in the tissue, standards were added for visualization purposes. Results here indicate that the radioactive compounds migrated together with the standard. By comparing the lipid spots shown by the charring procedure with those in the autoradiograms, it is possible to observe radioactive spots that cannot be visualized by the charring procedure.

Important Tips for Successful Separation of the Phospholipids

1. In our laboratory, all HPTLC separations are carried out in rectangular glass solvent tanks (27 cm long × 13 cm high × 7 cm wide) obtained from Alltech Associates, Inc. (Deerfield, IL). Each tank can accom-

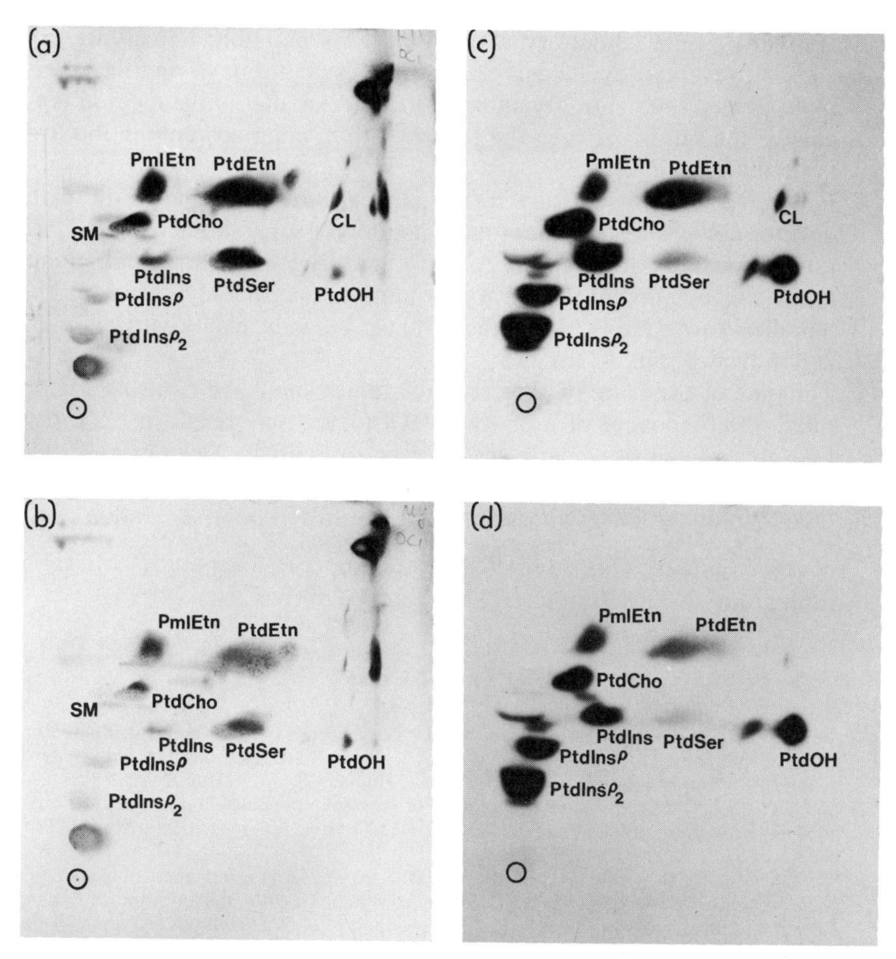

FIG. 1. Separation of phospholipids of **(a)** brain homogenate and **(b)** myelin by the two-dimensional HPTLC procedure as described in the text. The brain samples were spiked with polyphosphoinositide standards (100 μg) prior to spotting the plates. After solvent development, lipids were visualized by charring with cupric acetate. Chromatograms to the right depict the autoradiograms of the ^{32}P-labelled phospholipoids in **(c)** brain homogenates and **(d)** myelin. PmlEtn, ethanolamine plasmalogens; PtdEtn, phosphatidylethanolamine; PtdIns, phosphatidylinositol; PtdCho, phosphatidylcholine; CL, cardiolipin; SM, sphingomyelin. For other abbreviations, see the text. (Reproduced with permission from Ref. 2.)

modate 12 10 cm × 10 cm HPTLC plates and the entire procedure can be finished within 3 h.

2. The silica gel HPTLC plates obtained from different sources may have slightly different properties and thus give rise to differences in the separation. We have consistently used the Silica Gel 60 HPTLC plates from E. Merck.

3. Humidity in the laboratory may affect the separation. Sometimes this can be prevented by storing plates in a desiccator or drying the plates in an oven at 90°C for 30 min prior to use. An alternative method is to change the ratios between the polar and non-polar solvents in the first system.
4. Failure to remove moisture from samples during spotting of the plate or after solvent development may also affect separations. In our laboratory, we found it necessary to use a hair drier with a strong airstream and the setting on hot. Under this condition, one should never point the hair drier towards any part of the plate but blow the plates with a fanning action from about 30 cm away.
5. The time of exposure of plates to HCl fumes should be controlled carefully. Overexposure of a plate to HCl fumes may result in excessive hydrolysis of fatty acids from other phospholipids. Deleting this step results in poor separation of PtdIns, PtdSer and PtdOH.
6. Overloading the plate with samples will result in poor separation.

Acknowledgments This research project is supported in part by USDHHS research grants NS20836 from NINCDS and AA06661 from NIAAA.

REFERENCES

1. Sun GY. Preparation and analysis of acyl and alkenyl groups of glycerophospholipids from brain subcellular membranes. In: Boulton AA, Baker GB, Horrocks LA, eds. *Neuromethods: lipids and related compounds*. vol. 7, Humana Press, Clifton, NJ. 1988;63–82.
2. Sun GY, Lin TN. Time course for labeling of brain membrane phosphoinositides and other phospholipids after intracerebral injection of [^{32}P]ATP. Evaluation by an improved HPTLC procedure. *Life Sci* 1989;44:689–696.
3. Sun GY, Huang HM, Kelleher JA, Stubbs EB, Sun AY. Marker enzymes, phospholipids and acyl group composition of a somal plasma membrane fraction isolated from rat cerebral cortex: a comparison with microsomes and synaptic plasma membranes. *Neurochem Int* 1988;12:69–77.

Methods in Inositide Research,
edited by Robin F. Irvine.

14

Separation of Novel Polyphosphoinositides

Kurt R. Auger, Leslie A. Serunian, and Lewis C. Cantley

*Department of Physiology, Tufts University School of Medicine,
Boston, Massachusetts 02111*

Within the past few years, this laboratory has identified and characterized a new phosphoinositide (PI) kinase activity that is intimately involved in cell growth and transformation. This kinase serves as a direct biochemical link between novel phosphoinositide pathways and certain growth factors, including the platelet-derived growth factor (PDGF) (1), insulin (2), colony-stimulating factor-1 (CSF-1) (3), and several oncogene products (e.g., the polyoma virus middle T/pp60$^{c\text{-}src}$ complex (4) v-*src* (5,6)) that have intrinsic or associated protein-tyrosine kinase activity (7,8).

This phosphoinositide kinase, PI 3-kinase, associates specifically with both anti-phosphotyrosine and anti-growth factor receptor immunoprecipitates from a variety of growth factor-stimulated cells (1,2,3) and with anti-middle T immunoprecipitates from polyoma virus middle T-transformed cells (9,10). PI 3-kinase phosphorylates the D-3 position of the inositol ring of phosphotidylinositol (PtdIns) to produce a novel product, phosphatidylinositol 3-phosphate (PtdIns(3)P) that is distinct from PtdIns(4)P, the predominant monophosphate form of cellular PtdIns (10). Moreover, recent work has demonstrated that PI 3-kinase activity phosphorylates not only PtdIns, but also PtdIns(4)P and phosphatidylinositol 4,5-bisphosphate (PtdIns(4,5)P_2) to generate two, additional novel phospholipids: PtdIns(3,4)P_2 and phosphatidylinositol 3,4,5-trisphosphate (PtdIns(3,4,5)P_3), respectively (1).

Detection of the novel polyphosphoinositides *in vivo* has been hampered by the fact that these lipids are present in relatively low abundance in almost all cells that have been investigated. Therefore, to detect these novel phospholipids in intact cells, it is necessary to incorporate high levels of radioisotope into the total cellular phosphoinositides.

Until recently, the inability of other laboratories to detect PtdIns(3)P, PtdIns(3,4)P_2 and PtdIns(3,4,5)P_3 *in vivo* has also been due to the failure of conventional methods to separate these novel isomers from the well-known and more abundant PIs. For example, PtdIns(3)*P* and PtIns(4)*P* co-migrate on nearly all one- and two-dimensional thin-layer chromatography (TLC) systems used; PtdIns(3,4)P_2 and PtdIns(4,5)P_2 also co-migrate on conventional TLC systems. In addition, high-performance liquid chromatography (HPLC) procedures that are commonly used to separate the deacylated phosphoinositides typically utilize salt gradients that cannot resolve the various deacylated phospholipid isomers. Here, we describe methods developed in our laboratory in collaboration with others to identify the novel products of PI 3-kinase activity *in vitro* and *in vivo* labelled with either [γ-^{32}P]ATP, *myo*-[^{3}H]inositol or [^{32}P]orthophosphate. The procedures presented are reproducibly used to examine and quantify these phospholipids.

IN VITRO PRODUCTION OF NOVEL POLYPHOSPHOINOSITIDES

Cells responsive to the aforementioned growth factors are made quiescent and then stimulated with the appropriate ligand to activate the intrinsic protein-tyrosine kinase activity of the growth factor receptor. Alternatively, transformed cells that are known to have PI 3-kinase associated with their respective oncoprotein are used. Transformed cells are best used when in an exponential growth state, in order to immunoprecipitate maximal lipid kinase activity. Plates of cells (usually 10 cm in diameter) are washed two times in ice-cold phosphate-buffered saline (PBS) and lysed in a standard lysis buffer (1% NP-40 (Sigma, St Louis, MO), 137 mM NaCl, 1 mM $MgCl_2$, 1 mM $CaCl_2$, 150 μM vanadate, 1 μg leupeptin/ml, 1 μg aprotinin/ml). After incubation on a rocker platform at 4°C for 15–20 min, the lysate is cleared by centrifugation at 12,000 *g* for 5 min at 4°C. Immunoprecipitation with anti-phosphotyrosine antibody or anti-growth factor receptor antibody is done at 4°C on a rocker platform for at least 2–3 h. Immune complexes are collected on Protein A-Sepharose CL 4B that has been pre-washed in 10 mM Tris·HCl (pH 7.5), 1% bovine serum albumin, and stored in PBS containing 0.02% (w/v) sodium azide. Immune complexes are washed three times with 1 ml of 1% NP-40 in PBS, two times with 0.5 M LiCl in 100 mM Tris·HCl (pH 7.5), and finally two times in TNE (10 mM Tris·HCl (pH 7.5) 100 mM NaCl, 1 mM EDTA). Phosphoinositide kinase assays are performed directly on the beads. Routinely, all three phospholipid substrates, namely, PtdIns (obtained from Avanti Polar Lipids, Birmingham, AL), PtdIns(4)*P* and PtdIns(4,5)P_2 (obtained from Sigma) are used. The final concentration of the PtdIns, PtdIns*P* and PtdInsP_2 substrates is 0.03 mg/ml each, in a carrier of phosphatidylserine at a final concentration of 0.1 mg/ml. Alternatively, a crude PI mixture purified from bovine brain can be obtained from Sigma and used at a final concentration of 0.2 mg/ml. This mixture contains all four phospholipids and thus eliminates the need to mix the individual lipids. The phospholipids stored in chloroform are placed in a 1.5 ml microfuge tube, dried under a stream of nitrogen gas, and resuspended in 10 mM Hepes (pH 7.5), 1 mM EGTA, prior to sonication. To ensure that the phospholipid substrates are resuspended adequately, a cup horn sonicator (Heat Systems Ultrasonics, Inc.) is used instead of a microprobe. After sonication (5–10 min at room temperature on setting 7), the lipid suspension, which is initially cloudy, becomes clear.

The PI 3-kinase reaction is usually done in a total volume of 50 μl and is initiated by the addition of 10–50 μCi of [γ^{32}P]ATP (Dupont NEN, 3000 mCi/mmol) in a carrier of 50 μM unlabelled ATP, 10 mM Mg^{2+} and 20 mM Hepes (pH 7.5) to the washed immune complexes that have been pre-incubated for 5 min with the phospholipid substrates. The enzyme reaction is incubated at room temperature for 5–10 min. After stopping the reaction with 80 μl of 1 M HCl and extracting the lipids with 160 μl of chloroform/methanol (1:1, v/v), the ^{32}P-labelled phospholipid products in the bottom organic phase are collected after a brief centrifugation and stored at −70°C until further use.

These samples are often analysed and resolved on TLC plates (MCB Reagents, Merck, Silica Gel 60, 0.2 mm thickness) that have been pre-coated with 1% (w/v) potassium oxalate and baked at 100°C for 30 min to 1 h immediately before use. Unlabelled phospholipid standards are run in parallel to monitor lipid migration and are visualized by exposure to iodine vapour.

In order to separate the highly phosphorylated PtdIns(3,4,5)P_3 from the radioactivity remaining near the origin and any [γ-^{32}P]ATP and [^{32}P]phosphate carried over with the organic phase during the lipid extraction, an acidic solvent system of *n*-propanol/2.0 M acetic acid (13:7, v/v) is used instead of the more commonly used $CHCl_3$/MeOH/2.5 M NH_4OH (9:7:2, by vol.) solvent system. To achieve maximum resolution of each of the phospholipids from each other and from the material close to the origin, the solvent is allowed to migrate nearly to the top of a 20 cm TLC plate, a process that is routinely accomplished in 5 to 6 h.

DEACYLATION OF PHOSPHOLIPIDS

An HPLC system with a strong anion exchange (SAX) column and a shallow salt gradient results in baseline chromatographic separation of the glycerophosphoinositol phosphates (Gro*P*Ins) derived from the novel polyphosphoinositides and from those derived from the conventional PIs. This HPLC system separates the Gro*P*Ins molecules on the basis of structural differences in their inositol head groups and requires the removal of fatty acyl groups (deacylation) from the phospholipids prior to analysis.

The deacylation procedure is performed as follows. Labelled phospholipids that have been produced as described above and extracted in chloroform are washed with an equal volume of methanol/0.1 M EDTA (1:0.9, v/v) and then placed in a glass, screw-capped scintillation vial (20 ml capacity), and dried under a stream of nitrogen gas. In a fume cupboard, 1.8 ml of methylamine reagent (42.8% of 25% (v/v) methylamine in H_2O, 45.7% (v/v) methanol, 11.5% (v/v) *n*-butanol, stored at 4°C) are added to dissolve and hydrolyze the dried lipids. The tightly capped vial is incubated at 53°C with constant shaking for 50 min. After this incubation, the contents of the vial are cooled to room temperature, transferred to a 2.0 ml microfuge tube, and dried *in vacuo*. To prevent methylamine vapour damage to the vacuum pump and freeze-drying apparatus, a flask containing concentrated sulphuric acid (approximately 200 ml) is placed in solid CO_2 and is attached to the roto-evaporator to serve as an "acid trap". Optimally, the acid is cooled until it becomes a slurry, but is not completely frozen. A second flask is attached to the acid trap containing NaOH pellets to sequester and neutralize any acid in the system. Distilled, deionized H_2O (1.5 ml) is added to the dried contents of the tube, the sulphuric acid trap is removed, and the tube contents are dried again under vacuum. The dried samples are resuspended in 2.0 ml of distilled, deionized H_2O, transferred to a glass test tube, and extracted two times with an equal volume of *n*-butanol/petroleum ether (b.p. 40–60°C)/ethyl formate (20:4:1, by vol.) to remove fatty acyl groups. After these extractions, the lower, aqueous phase is dried *in vacuo* by roto-evaporation and stored at −70°C until analysis by HPLC.

DEACYLATION OF TLC-PURIFIED PHOSPHOLIPIDS

In instances when phospholipids are separated by TLC, HPLC analysis can still be performed after deacylation of the phospholipids *in situ*. To this end, all TLC-purified lipids are deacylated in the following manner. The region of the TLC plate that contains a particular phospholipid spot is carefully excised, placed in a glass screw-capped scintillation vial, and treated with methylamine reagent exactly as described above. We have found that it is not necessary to scrape the silica off the plate (unless glass plates are used). Oxalate treatment of the TLC plate obviates the need to wash the organic extract with methanol/0.1 M EDTA to remove divalent cations. After the incubation in methylamine, the solution is removed from the vial and placed in a 2.0 ml microfuge tube. If part of the silica gel from the TLC plate has chipped off and been carried over with the solution, a brief centrifugation is done and the supernatant removed to a fresh tube. The remaining steps in the deacylation procedure are then completed as detailed above. The TLC-purified and deacylated glycerophosphoinositides are then analysed by HPLC (see below).

HPLC ANALYSIS OF DEACYLATED PHOSPHOLIPIDS (GLYCEROPHOSPHOINOSITIDES)

Baseline separation of all of the glycerophosphoinositides is achieved using an HPLC high-resolution 5 μM Partisphere SAX column (Whatman) and a shallow, discontinuous salt gradient. Dried, deacylated samples are resuspended in 0.1–0.5 ml of 10 mM $(NH_4)_2HPO_4$ (adjusted to pH 3.8 with H_2PO_4) and applied to the column for analysis.

Routinely, PIs labelled with a different radioisotope and deacylated exactly as described above are co-injected with each sample to serve as standards. For example, ^{32}P-labelled polyphosphoinositides produced *in vitro* by immunoprecipitated PI 3-kinase activity are deacylated and applied to the HPLC column simultaneously with commercially available [^{3}H]PtdIns(4)*P* and [^{3}H]PtdIns(4,5)P_2 (Dupont NEN) deacylated as described. The non-radioactive nucleotides ADP and ATP are also added to allow monitoring with an ultraviolet (UV) detector to evaluate the reproducibility of elution times among all of the HPLC runs. The samples are loaded on to the HPLC column using a Hamilton syringe with a blunt-ended needle.

The HPLC column is equilibrated with H_2O prior to sample loading and is eluted with a discontinuous gradient up to 1 M $(NH_4)_2HPO_4{\cdot}H_2PO_4$ (pH 3.8) at a flow rate of 1 ml/min. The gradient is established from 0 to 1.0 M $(NH_4)_2HPO_4{\cdot}H_2PO_4$ over 115 min. To develop the shallow gradient required for these separations, dual pumps are used: pump A contains H_2O; pump B contains 1.0 M $(NH_4)_2HPO_4{\cdot}H_2PO_4$. At the start, B is run at 0% for 10 min, to 25% B with a duration of 60 min, and then to 100% B with a duration of 45 min. For an alternative gradient elution that separates the novel isomers with better baseline distinction between Gro*P*InsP_2s, the following gradient is used; B is run at 0% for 5 min, to 15% B with a duration of 55 min, isocratic elution for 20 min, and then to 100% B over 25 min (Fig. 1). The column is washed at 100% B for 10 min and then re-equilibrated with water for the next analysis.

The eluate from the HPLC column is connected first to an on-line UV monitor and then directly to an on-line continuous flow liquid scintillation detector (Flo-One/Beta CT; Radiomatic Instruments, FL) that can monitor and quantify two different radioisotopes simultaneously. Data are collected and stored by the radioactive flow detector and then transmitted to a Macintosh SE for further analysis.

In the event that an automatic radioactive flow detector is unavailable, it is possible to analyse radiolabelled phospholipid samples by attaching a fraction collector to the HPLC column and collecting individual eluate fractions (approximately 100–500 μl) at 5–30 s intervals over the entire gradient or in the region of interest. Scintillation vials are used for collection instead of tubes to avoid transferring samples. The appropriate volumes of water and

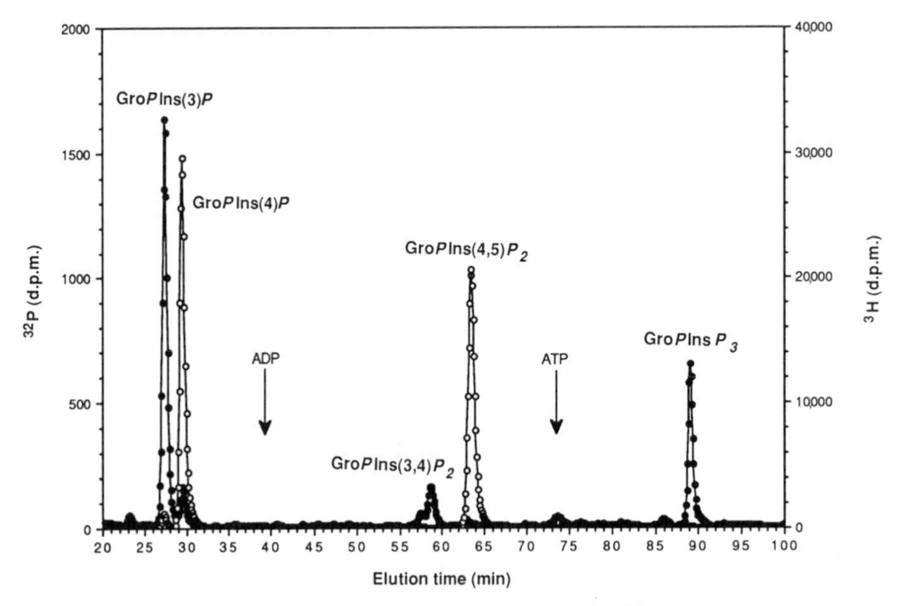

FIG. 1. HPLC separation of deacylated phosphoinositides. [^{32}P]phosphoinositides (●) generated *in vitro* from anti-phosphotyrosine immunoprecipitates of PDGF-stimulated cells were deacylated and analysed as described. [^{3}H]Inositol-labelled standards (○), ADP and ATP were added to the samples prior to analysis. This tracing represents a typical analysis, using the second gradient described, which increases the separation of the Gro*P*InsP_2s. The data were transferred to a Macintosh SE for graphing.

scintillation fluid are added to the eluate fractions in each vial, and the vials are capped, vortexed and counted in a scintillation counter. Although this alternative method is quite reliable, it is very labour-intensive and slow because of the number of fractions that must be counted.

MYO-[^{3}H]INOSITOL- AND INORGANIC ^{32}P-LABELLING OF CULTURED CELLS FOR *IN VIVO* DETECTION OF NOVEL POLYPHOSPHOINOSITIDES

Cells are plated in 10-cm tissue culture plates in standard growth medium and placed at 37°C in a humidified atmosphere containing 5% (v/v) CO_2. After the cells have attached to the substratum, 5 to 6 ml of inositol-free Dulbecco's Modified Eagle's Medium (DMEM) (GIBCO) supplemented to contain 10% (v/v) dialysed foetal bovine serum (M_r cutoff 3400) and 10–20 μCi/ml of *myo*-[^{3}H]inositol ([^{3}H]Ins, ARC, 15 Ci/mmol) are added to each plate. Cells are incubated for 24 to 96 h in this medium at 37°C and then are placed in serum-free conditions: 5 ml of inositol-free DMEM and 10–20 μCi/ml of [^{3}H]Ins. Cells are incubated for an additional 12 to 72 h under serum-free conditions in the presence of [^{3}H]Ins to induce quiescence. In order to ensure that the less abundant novel polyphosphoinositides are adequately labelled *in vivo*, radioisotope should initially be added while the cells are still in exponential growth phase. In addition, by allowing the cells to reach quiescence in the presence of the label, a significant amount of radioactivity remains incorporated into the inositol-containing cellular phospholipids.

Cells are labelled with inorganic [^{32}P]orthophosphate in a manner similar to that described above for [^{3}H]inositol labelling. However, for [^{32}P]orthophosphate labelling, quiescent cells are washed from their initial plating medium and placed in 5 ml of phosphate-free DMEM (GIBCO) supplemented to contain 0.5% (w/v) bovine serum albumin and 0.5–2 mCi of carrier-free [^{32}P]orthophosphate (Dupont NEN). The plates are incubated under a protective Plexiglass shield for 2–4 h at 37°C in a humidified atmosphere containing 5% (v/v) CO_2.

For mitogen stimulation, recombinant or purified growth factor is added to the cell culture medium at an optimal mitogenic concentration for a specific period of time prior to the labelled cells being washed and harvested and the lipids extracted. With the exception of time course studies of growth factor stimulation, mitogen is usually added to plates of cells for 5–15 min to detect maximum elevation of the novel polyphosphoinositides. Cells are incubated with mitogen at 37°C in a humidified incubator containing 5% CO_2.

To extract cellular lipids, individual plates of cells are washed three times with ice-cold PBS to remove unincorporated radioisotope. Cells are harvested with a cell scraper in 750 μl of methanol/1 M HCl (1:1, v/v) and placed in a 1.5 ml microfuge tube. After vigorous vortexing, 380 μl of chloroform is added and the tube is vortexed and incubated on a rocker platform for 15 min at room temperature. The samples are centrifuged briefly to separate the phases. The upper aqueous phase is carefully removed, placed in a clean microfuge tube, and dried by roto-evaporation *in vacuo*. Approximately 1.0 ml of distilled, deionized H_2O is then added to the tube and the contents are dried again. The dried aqueous sample is stored at −70°C until further use. In contrast, the organic phase and the material at the interface are extracted two times with an equal volume of MeOH/0.1 M EDTA (1:0.9, v/v) to remove traces of divalent cations (11). The entire organic phase is carefully removed, placed in a clean microfuge tube, and stored at −70°C under nitrogen gas until deacylation (described above) or TLC analysis.

We have found it necessary to load a few million d.p.m. of [^{3}H]Ins-labelled deacylated lipids in order to detect the novel isomers as they constitute only a small fraction of the total PtdInsP_2 in the cell. In addition, PtdIns(3,4)P_2 and PtdIns(3,4,5)P_3 are not present in most cells until they are induced by mitogen or when the cells are proliferating. We periodically detect another [^{3}H]inositol-labelled compound eluting approximately one and a half minutes prior to the deacylation product of PtdIns(3,4)P_2 that has tentatively been identified as deacylated PtdIns(3,5)P_2 (1).

Acknowledgment This work was supported by grant GM36624 from the National Institutes of Health.

REFERENCES

1. Auger KR, Serunian LA, Soltoff SP, Libby P, Cantley LC. PDGF-dependent tyrosine phosphorylation stimulates production of novel polyphosphoinositides in intact cells. *Cell* 1989;57:167–175.
2. Ruderman NB, Kapeller R, White MF, Cantley LC. Activation of phosphatidylinositol 3-kinase by insulin. *Proc Natl Acad Sci USA* 1990;87:1411–1415.
3. Varticovski L, Drucker B, Morrison D, Cantley L, Roberts T. CSF-1 receptor associates with and activates phosphatidylinositol 3-kinase. *Nature* 1989;342:699–702.
4. Kaplan DR, Whitman M, Schaffhausen B, et al. Phosphatidylinositol metabolism and polyoma-mediated transformation. *Proc Natl Acad Sci USA* 1986;83:3624–3628.
5. Sugimoto Y, Whitman M, Cantley LC, Erikson RL. Evidence that the Rous sarcoma transforming gene product phosphorylates phosphatidylinositol and diacylglycerol. *Proc Natl Acad Sci USA* 1984;81:2117–2121.
6. Fukui Y, Hanafusa H. Phosphatidylinositol kinase activity associates with viral $p60^{src}$ protein. *Mol Cell Biol* 1989;9:1651–1658.
7. Kaplan DR, Whitman M, Schaffhausen B, et al. Common elements in growth factor stimulation and oncogenic transformation: 85 KD phosphoprotein and phosphatidylinositol kinase activity. *Cell* 1987;50:1021–1029.
8. Whitman M, Cantley L. Phosphoinositide metabolism and the control of cell proliferation. *Biochim Biophys Acta* 1988;948:327–344.
9. Whitman M, Kaplan DR, Schaffhausen B, Cantley L, Roberts TM. Association of phosphatidylinositol kinase activity with polyoma middle-T competent for transformation. *Nature* 1985;315:239–242.
10. Whitman M, Downes CP, Keller M, Keller T, Cantley L. Type I phosphatidylinositol kinase makes a novel inositol phospholipid, phosphatidylinositol-3-phosphate. *Nature* 1988; 332:644–646.
11. Clarke NG, Dawson RMC. Alkaline $O \rightarrow N$ transacylation. *Biochem J* 1981;195:301–306.

Methods in Inositide Research,
edited by Robin F. Irvine.
Raven Press, Ltd., New York © 1990

15

Quantification of Polyphosphoinositides by Acetylation with [^{3}H]Acetic Anhydride

Janet M. Stein*, Gerry Smith**, and J. Paul Luzio*

**Department of Clinical Biochemistry, University of Cambridge, Addenbrooke's Hospital, Cambridge CB2 2QR, UK and*
***Department of Biochemistry, University of Cambridge, Cambridge CB2 1QW, UK*

Quantification of diacylglycerol by acetylation of the hydroxyl group with [^{3}H]acetic anhydride was described by Banschbach et al. (1). Since acetic anhydride can react with either hydroxyl groups or amino groups we have developed a dual label method for quantification of cholesterol and phospholipids (2). The principle of the method is to acetylate the lipid and then isolate the lipid acetate from unchanged lipid by thin-layer chromatography (TLC). The addition, before the acetylation, of a sample of lipid labelled with a second isotope (^{14}C or ^{32}P is convenient) allows a ^{3}H/^{14}C or ^{3}H/^{32}P ratio to be obtained for the acetylated lipid; by comparison with the ratio obtained for a set of lipid standards treated in identical fashion, the amount of lipid in the unknown sample can be calculated. This value (a ratio) is not affected by losses in extraction and processing, counting efficiencies, or exactness of specific activity because the addition of a known number of counts of the ^{14}C- or ^{32}P-labelled lipid to both unknown and standard samples at the start of the procedure means that the initial concentration of the lipid present may be determined. The method offers high sensitivity, since it can assay amounts of polyphosphoinositide of less than 20 nmol and contrasts with lipid analyses in which results are presented as percentage recovered lipid.

We describe a method for acetylation of phosphatidylinositol phosphate (PtdIns(4)P) and phosphatidylinositol 4,5-bisphosphate (PtdIns(4,5)P_2) with [^{3}H]acetic anhydride and for separation of the products from each other and also from unchanged starting material. At present we have not found any commercial source of [^{14}C]PtdIns(4)P or [^{14}C]PtdIns(4,5)P_2 and so have prepared in this laboratory [^{32}P]PtdIns(4)P and [^{32}P]PtdIns(4,5)P_2 to use as the second labelled lipid for the assay.

MATERIALS AND METHODS

N,N-Dimethylformamide (DMF), acetic anhydride 98% pure, and *N,N*-diisopropylethylamine are from Aldrich Chemical Co. Ltd, Gillingham, Dorset, UK, 4-dimethylaminopyridine is from Kodak Ltd, Kirkby, Lancs., UK. PtdIns(4)*P* and PtdIns(4,5)P_2, both 98% pure, and crude bovine brain polyphosphoinositides are from Sigma Chemical Co., Poole, Dorset, UK. Pure phosphatidylcholine (PtdCho) (dioleoylglycerophosphocholine) is synthesized as described by Smith and Stein (3). [^{3}H]acetic anhydride supplied at 50 mCi/mmol is from NEN Research Products, Du Pont (UK) Ltd, Stevenage, Herts., UK. [^{32}P]P_i (40 mCi/ml) is from Amersham International, Amersham, Bucks, UK. X-ray film (Cronex 4) is from Dupont (UK) Ltd, Wedgewood Way, Stevenage, Herts., UK.

Oxalate Impregnated Plates

These are prepared from Polygram Sil N-HR precoated plastic sheets (supplied by Camlab, Nuffield Rd, Cambridge, UK) by soaking in a solution of 1.3% (w/v) potassium oxalate (pH 7.5), 1 mM EGTA, in water/methanol (3:2, v/v), then air-dried. Before use, the plates are activated by heating at 110°C for 30 min.

Bulk Polyphosphoinositide Acetate Preparation

For use as a carrier lipid, this is prepared from a crude brain polyphosphoinositide mixture (Sigma) essentially as detailed below in Experimental Procedure, but without radioactivity. The lipid (1 mg) is dried, DMF (400 μl) containing a trace of 4-dimethylaminopyridine, together with diisopropylethylamine (400 μl), and unlabelled acetic anhydride (100 μl) are added, the mixture is incubated for 24 h at 37°C and the lipid acetates are purified by an acid Folch extraction (see Experimental Procedure).

[^{32}P]PtdIns(4)*P* and [^{32}P]PtdIns(4,5)P_2

These are prepared from labelled ghosts made from human red cells incubated with [^{32}P]P_i as described by Allan and Michell (4). Red cells prepared from fresh human blood are washed twice in 150 mM NaCl, 5 mM glucose, 10 mM Tris·HCl (pH 7.4), and then twice in Krebs/Hepes/glucose (5) containing 0.1% (w/v) bovine serum albumin. The washed red cells (25 ml) are resuspended in the same medium, with added [^{32}P]P_i (1 mCi) and incubated overnight at 37°C. The red cells are isolated by centrifugation, and the pellet lysed with 14 volumes of 10 mM Tris·HCl (pH 7.4), 1 mM EGTA, 0.5 mM

EDTA. The red cell ghosts are first extracted with a neutral chloroform/methanol solvent system (6) to remove neutral phospholipids. Polyphosphoinositides are then obtained by extracting the ghosts with an acidified chloroform/methanol solvent system (7) that includes several washes of the lipid-containing organic phase to remove contaminating free [^{32}P]P_i. We have found that, in addition to [^{32}P]PtdIns(4)*P* and [^{32}P]PtdIns(4,5)P_2, the preparations also contain ^{32}P-labelled phosphatidic acid (PtdOH) (see Figs. 1 and 3 as reported) (4). Since no acetylated derivative can be formed from PtdOH (no hydroxyl or amino group to acetylate), the presence of PtdOH does not interfere with the assay and can be tolerated provided a TLC system is selected such that PtdOH does not overlap PtdIns(4)*P* acetate or PtdIns(4,5)P_2 acetate.

[^{3}H]Acetic Anhydride

This is supplied in a sealed ampoule and is adjusted to an appropriate specific activity (of approximately 8 mCi/mmol) by addition (with caution, in a fume cupboard) of a solution containing a known amount of acetic anhydride/ml toluene, to give 20 μCi and 2.5 μmol of acetic anhydride in 10 μl. This solution is stable for several months, provided H_2O is rigorously excluded. We keep [^{3}H]acetic anhydride in a glass-stoppered tube in a screw top container over dried silica gel at 4°C. If necessary this stock solution can be diluted to a lower specific activity by adding more acetic anhydride solution.

Estimation of Radioactivity

Scanning of radioactivity on TLC plates is by a Berthold LB 2842 TLC linear analyser. Efficiency of counting is 0.08% for ^{3}H and 30% for ^{32}P radioactivity.

EXPERIMENTAL PROCEDURE

The lipid-containing sample to be quantified (e.g., an acidified chloroform/methanol extract of membranes), and a set of standards containing known amounts (say 2–20 nmol) of unlabelled PtdIns(4)*P* or PtdIns(4,5)P_2 are set up and an identical number of c.p.m. of the ^{32}P-labelled lipids added to each sample, together with a small amount (say 5 μg) of pure PtdCho in order to prevent absorption of the highly polar polyphosphoinositides to the walls of the glass container. (Since PtdCho cannot be acetylated because it has no hydroxyl or amino groups this need not be measured accurately.) The mixture is dried down under nitrogen gas and then H_2O removed by evaporating for 1–2 h on a high-vacuum pump. Final removal of H_2O is facilitated by adding approximately 100 μl of DMF to each sample and then evaporating again on the vacuum pump overnight. DMF (40 μl), containing approx. 0.5 mg 4-dimethylaminopyridine/ml is added, together with *N,N*-diisopropylethylamine (20 μl), and finally [^{3}H]acetic anhydride (10–20 μl), containing 20 μCi and 2–5 μmol acetic anhydride/10 μl. The samples are sealed, then sonicated for 5 min in a bath sonicator to help the dispersion of the polyphosphoinositides. The sealed samples are enclosed in a screw top jar over freshly dried silica gel to exclude H_2O from the reaction, then incubated at 37°C for 8 h. A laboratory incubator is convenient for this.

The reaction is stopped with H_2O (100 μl) followed by occasional Whirlimixing for 15 min to hydrolyse the [^{3}H]acetic anhydride remaining. *After* this (i.e. after all of the unreacted acetic anhydride has been hydrolysed), 5–10 μg of unlabelled carrier polyphosphoinositide acetates (prepared as in Materials and Methods, above) are added to each sample, to aid extraction and TLC of the lipids, followed by 3 ml of chloroform/methanol (2:1, v/v) and then 600 μl of 0.6 M HCl to create two phases. The upper phase is discarded and the lower phase washed once with Folch (8) upper phase chloroform/methanol/0.6 M HCl (3:47:48, by vol.) and twice with chloroform/methanol/0.01 M HCl (3:47:48, by vol.) This removes free [^{3}H]acetic acid from the lipid acetates, and in addition the acid Folch extraction ensures that the lipid acetates are kept in protonated form and so partition into the organic phase. After this a few drops of methanol are added to each sample to clarify the organic phase. The samples are dried under a stream of nitrogen, followed by 1 or 2 h on a vacuum pump.

TLC is carried out on oxalate-impregnated plates (see Materials and Methods, above), activated at 110°C for 30 min before use. The solvent is chloroform/methanol/4 M ammonia (60:40:10, by vol.).

Detection of Spots and Measurement of Radioactivity

There are two methods:

1. Autoradiography of TLC plates is carried out to locate the labelled lipids and then the spots cut out and counted in a scintillation counter. We do not find it necessary to scrape off the plate but count it complete with plastic backing. The counting efficiency is improved if the samples are left to stand in scintillation mixture for about 24 h before counting.
2. A Berthold automatic TLC-linear analyser may be used. This measures total (^{32}P or ^{14}C + ^{3}H) radioactivity when the aperture is on open window setting, and ^{32}P, or ^{14}C, radioactivity with the aperture reduced using a polythene sheet (1.25 mg/cm^2).

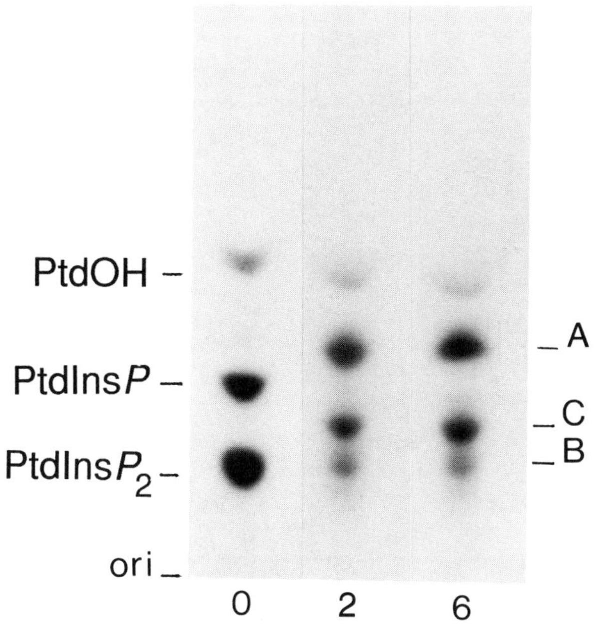

FIG. 1. Acetylation of ^{32}P-labelled polyphosphoinositides. The samples contained ^{32}P-labelled polyphosphoinositides, approximately 10,000 c.p.m. (see Materials and Methods) and approximately 5 μg of pure PtdCho. The samples were thoroughly dried. DMF (40 μl) containing 0.5 mg 4-dimethylaminopyridine/ml, diisopropylethylamine (20 μl), and unlabelled acetic anhydride (10 μl) were added; the samples were sonicated, then incubated at 50°C for 0, 2 and 6 h. The acetylated products were extracted and separated by TLC as in Experimental Procedures, and spots located by autoradiography. Abbreviations: PtdOH, phosphatidic acid; PtdIns(4)P, phosphatidylinositol 4-phosphate; PtdIns(4,5)P_2, phosphatidylinositol 4,5-bisphosphate; ori, origin.

RESULTS AND DISCUSSION

For polyphosphoinositide acetylation we use the dipolar aprotic solvent DMF to prevent inverted micelle formation and consequent exclusion of non-polar acetic anhydride from the reaction site. We also routinely include in the mixture diisopropylethylamine, an apolar hindered base, to ion-pair with the phosphate groups, and 4-dimethylaminopyridine as a catalyst.

Acetylation of PtdIns(4)P in this system gives PtdIns(4)P acetate (A in the TLC autoradiograph in Fig. 1 and in the scans of TLC plates of ^{32}P-labelled polyphosphoinositides labelled with [^{3}H]acetic anhydride in Figs. 2(a) and (b)), and is complete by 3 h at 37°C (Fig. 3(a)). PtdIns(4,5)P_2 is less easily acetylated, possibly because the high polarity of PtdIns(4,5)P_2 is not favourable for conversion to a less polar acetylated form, and in preliminary experiments we found little PtdIns(4,5)P_2 acetate formation in the absence

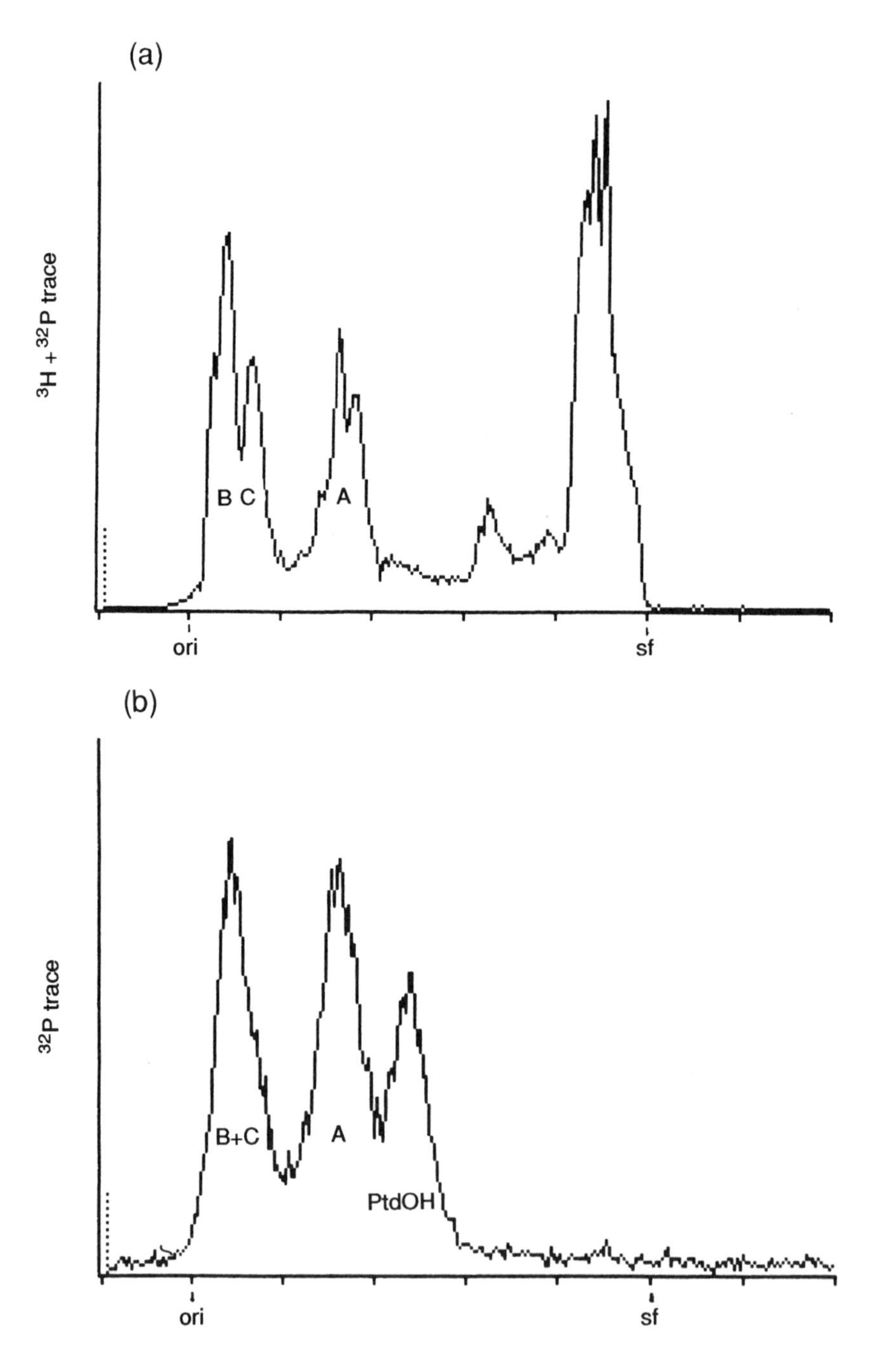

FIG. 2. Scanner profile of TLC separation of PtdIns(4)*P* acetate and PtdIns(4,5)P_2 acetate. A preparation of ^{32}P-labelled polyphosphoinositides containing approximately 4000 c.p.m. (see Materials and Methods) was acetylated for 8 h at 37°C with [^{3}H]acetic anhydride. Other conditions were as in Experimental Procedure. **(a)** Total radioactivity (^{32}P + ^{3}H)

of DMF. The main products of acetylation of PtdIns(4,5)P_2 run on TLC as two adjacent spots (B and C in Fig. 1). Examination of the scan of total counts of this product after acetylation of ^{32}P-labelled polyphosphoinositides with [^{3}H]acetic anhydride reveals two contiguous peaks (Fig. 2(a), peaks B and C), although B and C cannot be separately resolved for the same scan when on the reduced setting for ^{32}P (Fig. 2(b)). We find that B and C have similar ^{3}H/^{32}P ratios in the dual label assay (data not shown). However, since the proportion of B varies from 5% to 50% of the total PtdIns(4,5)P_2 acetate we do not use data from B in the assay. It is probable that C is fully acetylated PtdIns(4,5)P_2 and the more polar B is partially acetylated PtdIns(4,5)P_2. Calculations of PtdIns(4,5)P_2 acetate (Fig. 4(c) and (d)) use data from C exclusively.

Prolonged acetylation of PtdIns(4,5)P_2 also yields minor products giving less or well-defined spots which have no simple relationship between ^{3}H/^{32}P ratio and amount of PtdIns(4,5)P_2, and one of which overlaps with the PtdIns(4)P acetate peak. These products have not been characterized further. Acetylation at a higher temperature (50°C) also does not increase the speed of acetylation or give increased formation of the PtdIns(4,5)P_2 acetates (data not shown). A time course of acetylation of PtdIns(4,5)P_2 at 37°C (Fig. 3(b)) shows that prolonging the time of acetylation for more than 8 h does not increase the degree of acetylation; accordingly we routinely acetylate for 5-8 h at 37°C.

In the dual label experiment, we obtain a linear relationship between the ^{3}H/^{32}P ratio and nanomoles of PtdIns(4)P acetate or PtdIns(4,5)P_2 acetate (Fig. 4(a)–(d)). Data collected via scintillation counting or via the scanner give similar plots and either method may be used to measure polyphosphoinositides in biological samples. Factors for success are:

1. High specific activity ^{32}P- or ^{14}C-labelled PtdIns(4)P and PtdIns(4,5)P_2. The ^{32}P- or ^{14}C-labelled compounds should be the highest specific activity possible to minimize [^{3}H]acetate incorporation into the second labelled material which is being added; an alternative way of preparing these two lipids at a high specific activity is described in Chapter 3.
2. High specific activity [^{3}H]acetic anhydride. The specific activity of the [^{3}H]acetic anhydride should be high enough to give finally ^{3}H c.p.m. $>10 \times$ ^{32}P c.p.m. or ^{14}C c.p.m. for dual channel scintillation counting and $>1000 \times$ ^{32}P c.p.m. or ^{14} C c.p.m. for dual label counting using the scanner. In addition, for samples containing higher amounts of PtdIns(4)P or PtdIns(4,5)P_2 (say 20-100 nmol) the amount of unlabelled acetic anhydride used per assay must be increased further in order to ensure that the acetylation reaction goes to completion (i.e., there must

trace; **(b)** ^{32}P radioactivity trace only. Abbreviations: ori, origin; sf, solvent front; A is PtdIns(4)P acetate peak; B and C are PtdIns(4,5)P_2 acetate peaks. The abcissa measures distance (in cm) from origin (ori) of TLC plate. The solvent front was 10 cm from the origin.

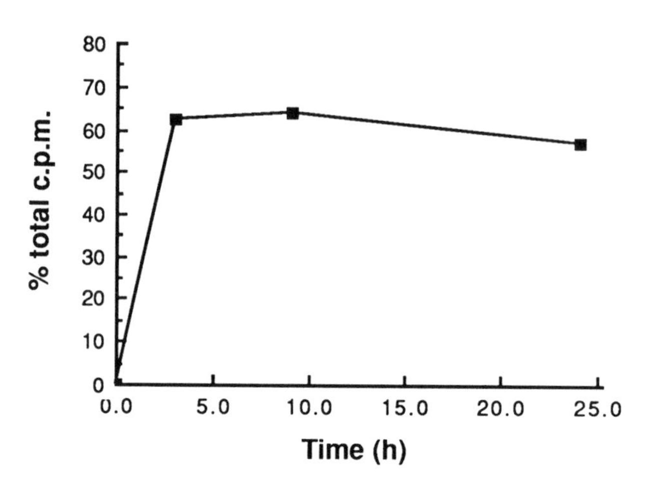

(b) Acetylation of PtdIns(4,5) P_2

% total c.p.m.

0 10 20 30 40 50

0.0 5.0 10.0 15.0 20.0 25.0

Time (h)

FIG. 3. Time course of acetylation of **(a)** PtdIns(4)*P* and **(b)** PtdIns(4,5)P_2. Samples of pure PtdIns(4)*P* (40 nmol) and PtdIns(4,5)P_2 (40 nmol), together with added PtdCho (approximately 5 μg)), were acetylated as in the legend to Fig. 1, except that 10 μl of [^{3}H]acetic anhydride (1.4 mCi/mmol) was used and incubation was at 37°C. The reaction products were separated by TLC and ^{3}H radioactivity quantified by scanner (see Experimental Procedure). There was negligible radioactivity in samples at 0 h.

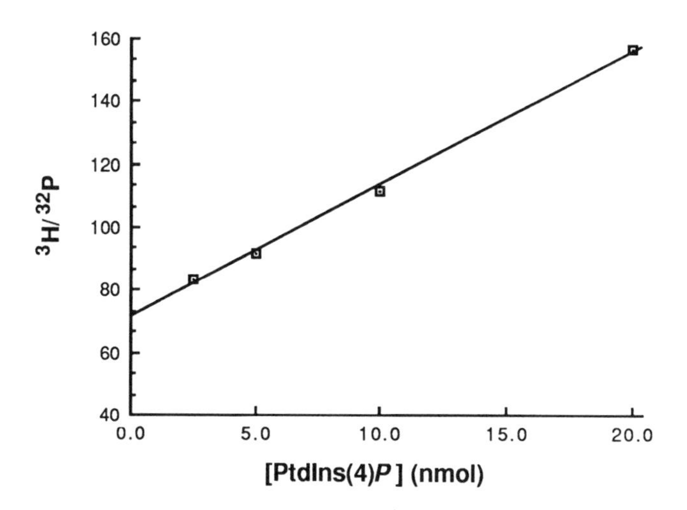

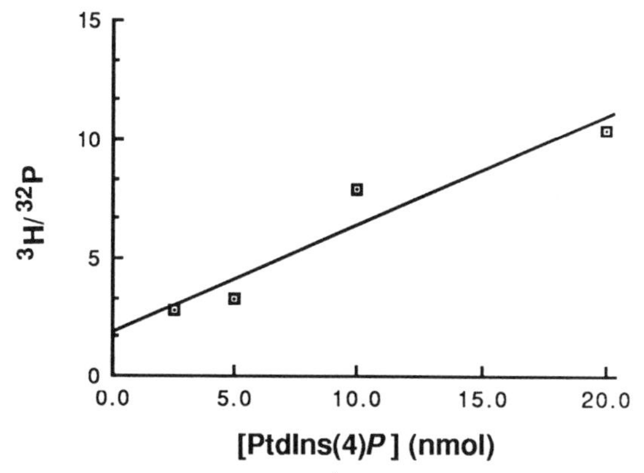

FIG. 4. PtdIns(4)*P* and PtdIns(4,5)P_2 standards. Each sample contained ^{32}P-labelled polyphosphoinositides (4000 c.p.m.), pure PtdCho (5 μg approximately), and a measured amount (2.5–20 nmol) of pure PtdIns(4)*P* or PtdIns(4,5)P_2. Other conditions were as in Experimental Procedure. ^{3}H/^{32}P ratios were plotted versus nmoles of PtdIns(4)*P* (**(a)** and **(b)**) and of PtdIns(4,5)P_2 **(c)** and **(d)**). Data for **(a)** and **(c)** were from scintillation counting and for **(b)** (**and (d)** from scanning. (*Figure continues.*)

(c) PtdIns(4,5)P_2 (from scintillation counting)

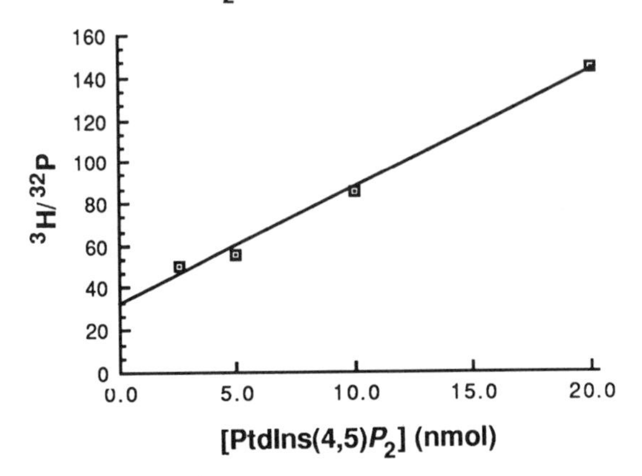

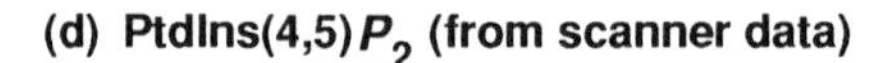
(d) PtdIns(4,5)P_2 (from scanner data)

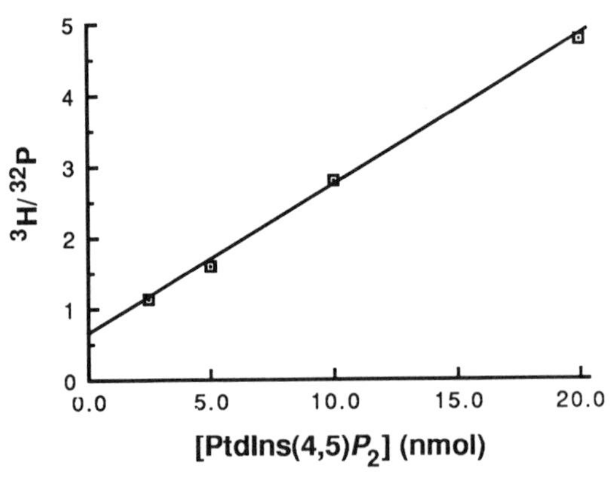

FIG. 4. *Continued.*

be sufficient acetic anhydride to drive the reaction forward). Therefore the [^{3}H]acetic anhydride as supplied (e.g., 50 mCi/mmol) must be diluted to an appropriate concentration with unlabelled acetic anhydride (see Materials and Methods). Otherwise the amount of radioactivity necessary for the reaction would be unacceptably hazardous.

3. Rigorous exclusion of H_2O at all stages of the acetylation.

The dual label method offers a method for quantifying very small amounts of lipid, with sensitivity limited only by the availability of high specific activity [^{3}H]acetic anhydride and of the added second ^{14}C- or ^{32}P-labelled lipid. Commercially available [^{14}C]PtdIns(4)*P* and [^{14}C]PtdIns(4,5)P_2 will improve the ease and speed of the assay. We are using a similar dual labelling assay for quantitative assay of cholesterol and phospholipids (2). We suggest that this approach could be adapted for quantification of the inositol phosphates and of glycolipids.

Acknowledgments This work was supported by a grant from the Medical Research Council. We thank Napp Laboratories, Cambridge Science Park, Cambridge, UK, for the loan of the Berthold TLC plate scanner.

REFERENCES

1. Banschbach MW, Geison RL, O'Brien JF. Use of [1-^{14}C]acetic anhydride to quantitate diglycerides: a new analytical procedure. *Anal Biochem* 1974;59:617–627.
2. Stein JM, Smith GA, Luzio JP. A new method for the quantitation of membrane lipids including phospholipids, polyphosphoinositides and cholesterol. Submitted for publication 1990.
3. Smith GA, Stein JM. *Techniques in the life sciences*, Biochemistry B4/1 *Techniques in lipid and membrane biochemistry*, part 1 *Lipid preparations* B406. Amsterdam: Elsevier/North-Holland Scientific Publishers Ltd, 1982.
4. Allan D, Michell RH. A calcium-activated polyphosphoinositide phosphodiesterase in the plasma membrane of human and rabbit erythrocytes. *Biochim Biophys Acta* 1978;508:277–286.
5. Newby AC, Luzio JP, Hales CN. The properties and extracellular location of 5′-nucleotidase of the rat fat-cell plasma membrane. *Biochem J* 1975;146:625–633.
6. Bligh EG, Dyer WJ. A rapid method of total lipid extraction and purification. *Can J Biochem Physiol* 1959;37:911–917.
7. Hauser G, Eichberg J, Gonzalez-Sastre F. Regional distribution of polyphosphoinositides in rat brain. *Biochim Biophys Acta* 1971;248:87–95.
8. Folch J, Lees M, Sloane-Stanley GH. A simple method for the isolation and purification of total lipids from animal tissues. *J Biol Chem* 1957;226:497–509.

Methods in Inositide Research,
edited by Robin F. Irvine.
Raven Press, Ltd., New York © 1990

16

Mass Measurement of Phosphatidylinositol 4-Phosphate and *sn*-1,2-Diacylglycerols

Nullin Divecha and Robin F. Irvine

Department of Biochemistry, AFRC Institute of Animal Physiology and Genetics Research, Babraham, Cambridge CB2 4AT, UK

Previous methods designed to determine the concentration of inositol-containing lipids have depended on the ability to radiolabel these lipids to equilibrium. Although this is relatively simple to perform in cell culture, it is more difficult in tissue slices, and, in the case of measuring changes in the mass of inositol lipids in the nucleus, very expensive (1).

Recently, monoclonal antibodies specific for phosphatidylinositol 4-phosphate (PtdIns(4)*P*) and another for phosphatidylinositol 4,5-bisphosphate (PtdIns(4,5)P_2) have been utilized to develop an enzyme-linked immunosorbent assay to measure mass changes of these lipids in Swiss mouse 3T3 cells after stimulation with growth factors (2). This method was demonstrated to be sensitive down to picomolar levels.

Another method employed for the measurement of lipids is based on the use of enzymes which specifically modify the molecule of interest, and in particular, lipid kinases, used with radiolabelled ATP, can detect mass levels in the picomolar range (for example, diacylglycerol (DAG) assays by an *Escherichia coli* DAG kinase (3)).

On the basis of the idea of using ^{32}P incorporation to quantify lipids and using a one-step purification to yield a "kinetically clean" PtdIns(4)*P* kinase (EC 2.7.1.68) and DAG kinase (EC 2.7.1.107) from rat brain cytosol, we have developed an assay for both PtdIns(4)*P* and DAG which is simple to perform, inexpensive, and is sensitive in the picomolar range.

METHODS

Purification of PtdIns(4)*P* Kinase

One rat brain is homogenized in 5 ml of buffer A (25 mM Tris-acetate, 25 mM ammonium acetate, 10 mM magnesium acetate, 0.32 M sucrose (pH 7.4)) and clarified by centrifugation at 50,000 *g*. The resulting supernatant (5 ml) is loaded on to a TSK-DEAE high-performance liquid chromatography (HPLC) column (BioRad) equilibrated in buffer A and the column developed with a linear gradient from buffer A to buffer A + 1 M ammonium acetate. Fractions (1 ml) are collected and assayed for PtdIns(4)*P* kinase activity as detailed below. Fractions containing enzyme activity are pooled, divided into 200-μl fractions and stored at −20°C. A typical profile of the elution of PtdIns(4)*P* kinase activity from the TSK-DEAE HPLC column is shown in Fig. 1(a) and this figure also shows the elution of DAG kinase. The PtdIns(4)*P* kinase elutes from this column at a salt concentration of approximately 0.5 M. This enzyme preparation was found to phosphorylate PtdIns(4)*P* linearly for up to 1 h and found to be stable at −20°C for two to three months (our unpublished data). One rat brain yields enough enzyme for approximately 200 PtdIns(4)*P* assays.

A satisfactory purification of PtdIns(4)*P* kinase can also be achieved in a single step by using a DEAE-Sepharose (Pharmacia 12 cm × 1.5 cm) column. In this case 16 rat brains are used and the column equilibrated in buffer A and developed using a linear gradient of buffer A to buffer A + 0.5 M ammonium acetate (200 ml of each). The chromatography profile obtained with this system is shown in Fig. 1(b). In this case PtdIns(4)*P* kinase eluted at 250 mM salt. PtdIns phosphodiesterase was found to elute in the flow through of the column, while DAG kinase was found to elute at a slightly lower salt concentration than did the PtdIns(4)*P* kinase. The important criterion for both these one-step purifications is that the enzymes catalysing the back reactions (PtdIns(4,5)P_2 phosphomonoesterase and phosphoinositidase C) are removed, so that the PtdIns(4)*P* kinase reaction is linear.

PtdIns(4)*P* Kinase Assay

PtdIns(4)*P* (1 nmol) is dried under nitrogen gas in an Eppendorf tube and sonicated at room temperature into 80 μl of PIPKIN buffer (50 mM Tris-acetate, 80 mM KCl, 10 mM magnesium acetate, 2 mM EGTA (pH 7.4)). A 20 μl portion of the fraction to be analysed is added and the reaction started by the addition of 100 μl of PIPKIN buffer containing ATP and 0.5 μCi [γ-^{32}P]ATP such that the final concentration in the reaction is 5 μM. The reaction is continued at 37°C for 30 min and stopped by the addition of 750 μl of chloroform/methanol/HCl (80:160:1, by vol.), followed by 250 μl of chloroform containing 0.5 μg P of Folch-fraction inositide mixture (prepared as in (4)) and finally 250 μl of 0.9% (w/v) NaCl. The mixture is vortexed and the phases split by centrifugation at room temperature in a Microcentaur for 2 min. The top phase is removed and the bottom phase washed three times with chloroform/methanol/1 M HCl (15:235:240, by vol.) The bottom phase is dried and counted by liquid scintillation counting techniques.

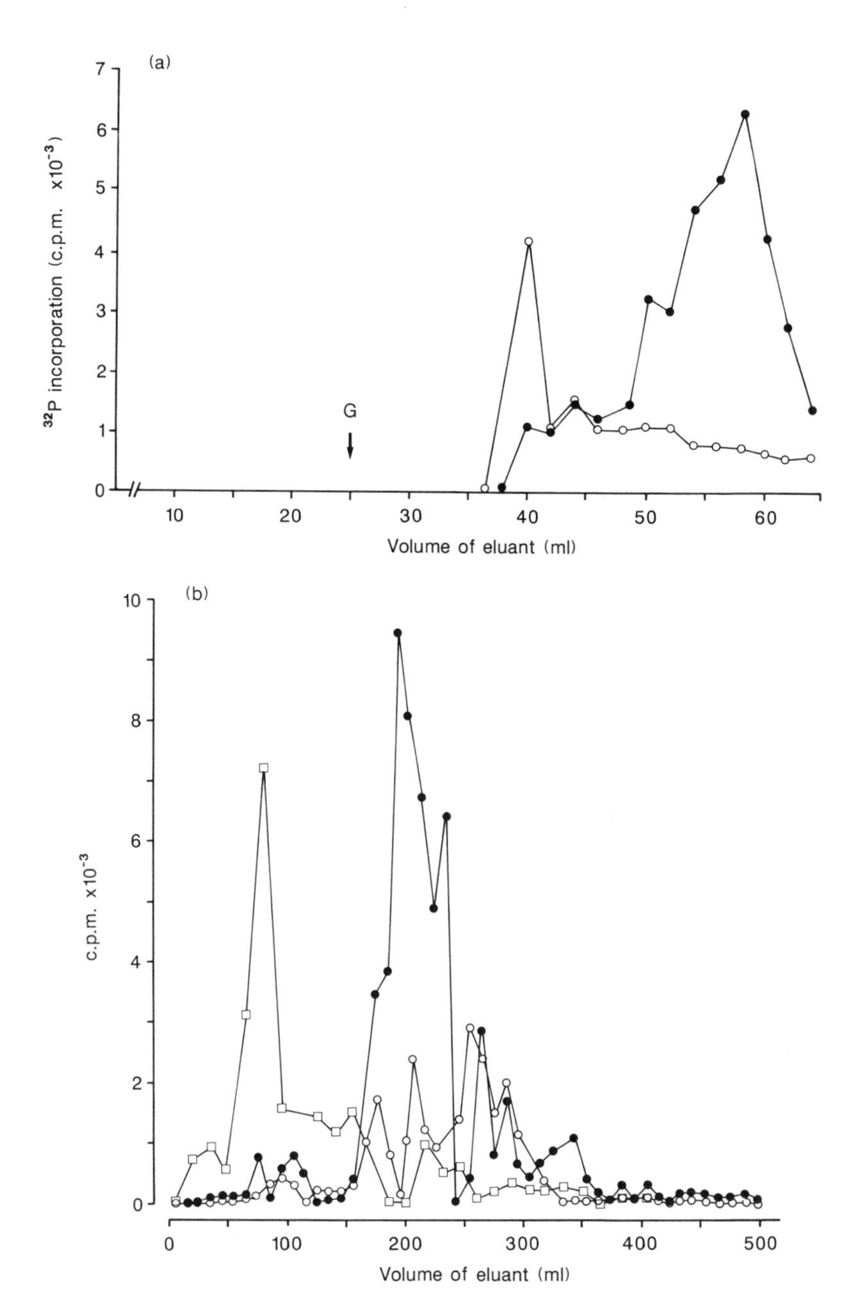

FIG. 1. (a) Elution profile of PtdIns(4)*P* kinase and DAG kinase from rat brain on a TSK-DEAE HPLC column. The experimental details are described in the text. (●) PtdIns(4)*P* kinase; (○) DAG kinase. G, start of gradient. **(b)** Elution profile of PtdIns(4)*P* kinase, PtdIns phosphodiesterase and DAG kinase from rat brain on DEAE-Sepharose. The experimental details are described in the text. (○) PtdIns(4)*P* kinase; (□) PtdIns phosphodiesterase; (●) DAG kinase.

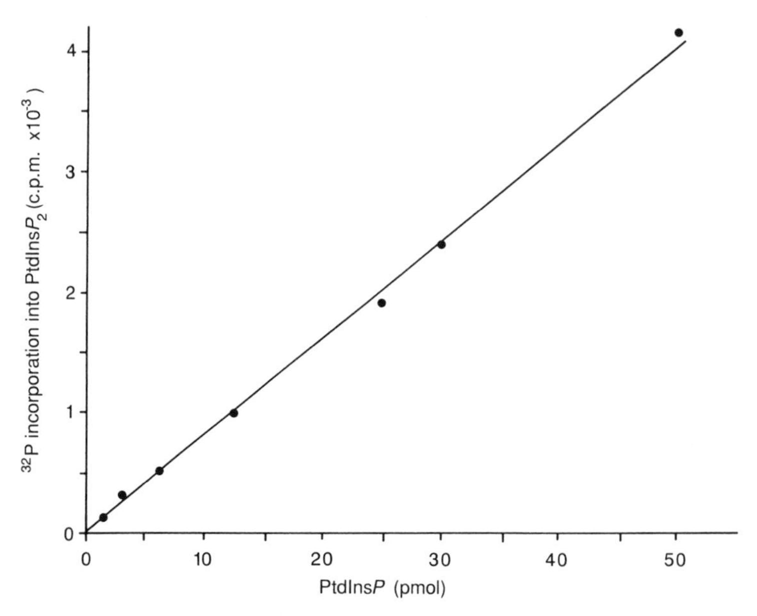

FIG. 2. Standard curve of PtdIns(4)*P* assay.

Mass Measurement of PtdIns(4)*P*

Total inositide lipids are isolated either from isolated nuclei or from cellular homogenates, by extraction with chloroform/methanol/HCl as detailed above, without the addition of the Folch carrier lipids, and dried under nitrogen in a 1.5 ml Eppendorf tube together with 50 nmol of phosphatidylethanolamine. PIPKIN buffer (100 μl) is added, and the sample is vortexed and then sonicated for 1 min in a bath sonicator. The reaction is started by the addition of 20 μl of pooled enzyme solution (from the DEAE columns as described), followed by 80 μl of PIPKIN buffer containing ATP and 1 μCi [^{32}P]ATP such that the final concentration of ATP is 5 μM. The reaction is continued for 1 h and the products extracted as described above. The dried lipids are then dissolved in 30 μl of chloroform and loaded on to oxalate (1%) sprayed thin-layer chromatography (TLC) plates, which have been activated by heating to 110°C in an oven. The TLC plates are developed in chloroform/methanol/concentrated ammonia/water (45:35:2:8, by vol.) (5) for 2 h. After autoradiography, the spots corresponding to PtdIns(4,5)P_2, as identified by co-chromatography with standards, are scraped off and their ^{32}P content is determined by scintillation counting. A standard curve can be constructed using known amounts of PtdIns(4)*P*.

As can be seen (Fig. 2), the phosphorylation of PtdIns(4)*P* to PtdIns(4,5)P_2 is linear between 3 pmol and 50 pmol. Addition of known amounts of

PtdIns(4)*P* before the extraction of lipids from the nuclei yielded essentially the same results as standards on their own, demonstrating that no inhibition of the enzyme by the extraction procedure or by other extractable lipids occurs. We are using this assay to look at changes in mass of PtdIns(4)*P* in the nuclei of Friend cells and preliminary results suggest that a two-fold increase in the concentration of PtdIns(4)*P* occurs during dimethylsulphoxide-induced differentiation (6).

A by-product of the partial purification of PtdIns(4)*P* kinase is the purification of a kinetically clean DAG kinase. We have used this to develop a mass assay for DAG based on the same principle as that of the assay of Priess et al. (3). The assay does not require exogenous cardiolipin for full activity, hence eliminating a major problem in current assay procedures. (Most commercially available cardiolipin contains substantial amounts of DAG.) The rat brain enzyme, however, appears to be inhibited by a contaminant in the DAG extracts themselves (see below). This can be readily removed by passing the DAG-containing lower phase of the extraction through a 1 ml silicic acid column. We have used this assay to look at the changes in mass of DAG on stimulation of Swiss mouse 3T3 cells with bombesin and this is described below.

Extraction of DAG from Swiss Mouse 3T3 Cells

Quiescent cells are stimulated with 10 nM bombesin as described (7) and the reaction stopped at the times indicated by the addition of 1 ml of ice-cold methanol. The cells are scraped and the wells washed with a further 0.5 ml and then 1.5 ml of chloroform. The phases are split by the addition of 1 ml of chloroform, 1 ml of 0.9% (w/v) NaCl followed by centrifugation at 1500 *g*. The lower phase is washed two times with 2 ml of chloroform/methanol/H_2O (15:235:245, by vol.) and the lower phase passed through a 1 ml silicic acid column equilibrated in chloroform. The flow through is collected, dried under nitrogen gas and assayed for DAG.

Mass Measurement of DAG

The dried lipid is dissolved by the addition of 20 μl of CHAPS (9.2 mg/ml H_2O; Sigma, St Louis, MO) and sonicated at room temperature for 15 s after the addition of 80 μl of PIPKIN buffer (see PtdIns(4)*P* kinase assay, above). The reaction is started by the addition of 20 μl of pooled DAG kinase enzyme followed by 80 μl of PIPKIN buffer containing ATP and 1 μCi [^{32}P]ATP such that the final concentration is 5 μM. The reaction is continued for 1 h at room temperature and the lipids extracted, separated and quantified as for the PtdIns(4)*P* assay. A standard curve can be constructed using known amounts of standards. As for the PtdIns(4)*P* assay, addition of known quan-

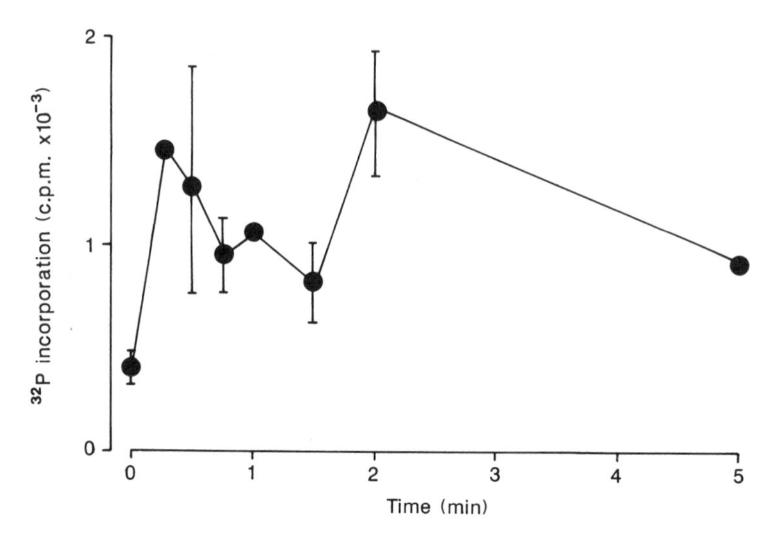

FIG. 3. Production of DAG in Swiss mouse 3T3 cells pretreated with phorbol myristoylacetate, and then stimulated with 10 nM bombesin. DAG was assayed by the method described in the text; incorporation into PtdOH is shown, because the cell numbers were not counted and so the mass of DAG per cell cannot be calculated.

tities of DAG added to the extraction of DAG from cells yield essentially the same results as standards on their own, as long as they were passed through the silicic acid column (data not shown).

Figure 3 shows the changes, on stimulation with 10 nM bombesin, in the DAG content of Swiss 3T3 cells after down-regulation of protein kinase C by long-term treatment with phorbol myristoylacetate (two days at 300 μM). As can be seen there is a three-fold increase in DAG on stimulation with bombesin, which is in agreement with the data published by Cook and coworkers (7). However, after 5 min the amount of DAG had returned to close to the baseline value. In control cells, stimulation with bombesin causes an initial increase in DAG followed by a sustained elevation (7). This sustained increase in DAG hs been suggested to be derived from either phospholipase C or D action on phosphatidylcholine and so it appears that protein kinase C may be involved with this effect.

The partial purification of DAG kinase and the PtdIns(4)*P* kinase from rat brain and the use of these enzymes in the assay of these lipids is simple and inexpensive, and both assays are sensitive in the lower picomolar range. Theoretically, it should be possible to increase the sensitivity of the assays down to the femtomolar level by further purification of the enzymes. This will be important if the assays are to be used to measure changes in nuclear lipids in 3T3 cells on stimulation with insulin-like growth factor 1 (8) and in Friend cells undergoing differentiation (6). The presence of PtdIns phosphodiesterase in the flow through from the DEAE-Sepharose column opens the possibility of also developing a coupled kinase assay for this lipid and this is currently being exploited in our laboratory.

Acknowledgments We thank Len Stephens for his invaluable help over in the development of these assays, Peter Smith for his initial work on the partial purification of PtdIns(4)*P* kinase and David Lander for purifying lipid standards. N.D. is supported by the Cancer Research Campaign.

REFERENCES

1. Horstman DA, Takemura H, Putney JW Jr. Formation and metabolism of [^{3}H]inositol phosphates in AR 42J pancreatoma cells. *J Biol Chem* 1988;263:15,297–15,303.
2. Matuoka K, Fukami N, Nakamishi O, Kawai S, Takenawa T. Mitogenesis in response to PDGF and bombesin abolished by microinjection of an antibody to PtdIns(4,5)*P*. *Science* 1988;239:640–643.
3. Preiss J, Loomis CR, Bishop WR, Stien R, Niedel JE, Bell RM. Quantitative measurement of *sn*-1,2-diacylglycerols present in platelets, hepatocytes, and *ras* and *sis* transformed normal rat kidney cells. *J Biol Chem* 1986;261:8597–8600.
4. Folch J. Complete fractionation of brain cephalin: isolation from it of phosphatidyl serine, phosphatidyl ethanolamine and diphosphoinositide. *J Biol Chem* 1949;177:497–505.
5. Irvine RF. The structures, metabolism and analysis of inositol lipids and inositol phosphates. In: Putney JW Jr, ed. *Phosphoinositides and receptor mechanisms*. New York: Alan R Liss, 1986;89–107.
6. Cocco L, Gilmour RS, Ognibene A, Letcher AJ, Manzoli FA, Irvine RF. Synthesis of polyphosphoinositides in the nuclei of Friend cells. *Biochem J* 1987;248:765–770.
7. Cook SJ, Palmer S, Plevin R, Wakelam MJO. Mass measurement of inositol 1,4,5-trisphosphate and *sn*-1,2-diacylglycerols in bombesin stimulated Swiss 3T3 mouse fibroblasts. *Biochem J* 1990;265:617–620.
8. Cocco L, Martelli AM, Gilmour GS, Ognibene A, Manzoli FA, Irvine RF. Rapid changes in phospholipid metabolism in the nuclei of Swiss 3T3 cells induced by the treatment of the cells with insulin-like growth factor 1. *Biochem Biophys Res Commun* 1988;154:1266–1272.

Methods in Inositide Research,
edited by Robin F. Irvine.
Raven Press, Ltd., New York © 1990

17

Mass Analysis of Molecular Species of Diradylglycerols

Arnis Kuksis and John J. Myher

Banting and Best Department of Medical Research, University of Toronto, Toronto, Ontario, Canada M5G 1L6

Natural glycerophospholipids, including the inositol phosphatides, are made up of molecular species containing different fatty acids, which occur in characteristic pairings (1). The diacylglycerol moieties may differ between phosphatidylinositol and its mono and diphospho derivatives, although molecular species of the polyphosphoinositol lipids have seldom been investigated (2,3). Furthermore, the molecular species of the phosphatidylinositol glycans may also contain alkylacyl glycerol moieties of varying chain length, in addition to the diacylglycerol moieties (4). The molecular species of these diradylglycerols are best separated, identified and quantified by chromatographic methods following their release from the parent compounds by phospholipase C (2,5), although a successful separation and identification of the common phosphatidylinositols has also been achieved by reversed-phase high-performance liquid chromatography (HPLC) of the intact molecules (6,7).

A comparative determination of the molecular species of the glycerophosphatides and free diacylglycerols in tissues is of great interest because a variety of agonist-stimulated events is associated with release of free diradylglycerols from inositol (8,9) and choline (10) phosphatides, and possibly from triacylglycerols (11). The identification of the molecular species of the released diacylglycerols (3,7,12) is helpful in establishing the origin (source) of these diradylglycerols, their cellular localization, and their role in cellular signalling and metabolism. In view of the occurrence of different isozymes of protein kinase C (13), which may show differential sensitivity to specific molecular structures (14), a knowledge of the composition of molecular species of both bound and free diradylglycerols is crucial. At present, there are several methods for determining the exact molecular identity of the diradylglycerols released in response to agonist stimulation, as well as the identity of those diradylglycerols normally occurring in cell membranes and organelles. Many of these analyses are performed almost as easily as fatty acid determinations.

In this chapter experimental details are presented for modern methods of analysis of bound and free diradylglycerols based on polar capillary gas–liquid chromatography (GLC), and reversed and chiral phase HPLC, including combinations with mass spectrometry. The chapter contains a thorough discussion of stereospecific analysis of enantiomeric diradylglycerols by chiral phase HPLC, which is necessary to distinguish between the *sn*-1,2-diradylglycerols arising from agonist-stimulated degradation of glycerophospholipids, and any *rac*-1,2-diacylglycerols arising from lipolysis of triacylglycerols.

ISOLATION AND PURIFICATION OF INOSITOL PHOSPHATIDES

Phosphatidylinositol (PtdIns) is the major inositol lipid in mammalian tissues, but appreciable quantities of the 4-phosphate (PtdIns*P*) and the 4,5-bisphosphate (PtdInsP_2) are also found in nervous tissue, adrenal medulla and kidney (15). Since post-mortem hydrolysis of PtdInsP_2 to PtdIns*P* and PtdIns is rapid, accurate analyses of the higher inositol phosphatides are difficult to obtain.

Inositol Lipids

A quantitative extraction of the polyphosphoinositol phospholipids requires acidification of the extraction medium, which is best performed following an initial removal of the neutral phospholipids with neutral solvents (16,17). (Direct extraction with acidic solvents leads to destruction of alkenylacylglycerolipids (16) that may be present.) The neutral solvents also extract PtdIns and some PtdIns*P*, which may be recovered by selective adsorption on neomycin columns (3,17,18, and see Chapter 1). The neutral solvent extracts may be combined with the acidic extracts after the latter have been neutralized (16). The individual phospholipid classes are usually resolved by normal phase thin-layer chromatography (TLC) and HPLC, although complete resolution of the inositol phosphatides may also be obtained on neomycin columns with careful selection of eluants (17,18, and Chapter 1). The inositol phosphatides are resolved by one-dimensional TLC (19) or high-performance thin-layer chromatography (HPTLC) (20) on silica gel impregnated with oxalic acid and by two-dimensional TLC (21, and see also Chapter 13). It should be noted that inositide carriers must be carefully selected, if mass analyses of molecular species are to be performed. If possible, use of carriers should be avoided. Likewise, iodine staining of TLC plates should be avoided because this can lead to sample decomposition. Fukami and Takenawa (22) have described mass analysis of PtdIns*P* and

PtdInsP_2 by TLC with immunostaining. The lipids are separated on a TLC polygram (Macherey and Nagel, FRG) and developed with chloroform/methanol/aqueous ammonia/H_2O (90:65:12:8, by vol.) The TLC plate is then soaked overnight in phosphate-buffered saline containing 3% (w/v) bovine serum albumin, 1% (w/v) polyvinylpyrrolidone 40. Afterwards the plate is treated for 2 h at room temperature with ascites of anti-PtdIns*P* and anti-PtdInsP_2 antibodies, followed by peroxidase-conjugated anti-mouse immunoglobins, and stained with a Konica staining kit (Konica, Tokyo) and measured by an image-processing system (AS Co., Nara, Japan).

Boyle et al. (23) have described a normal phase HPLC procedure for separating polyphosphoinositides on hydroxyapatite. A solvent system consisting of tetrahydrofuran, ethanol, water, with a gradient of triethylamine phosphate ranging from 1 to 100 mM is used to obtain a complete resolution of standard PtdIns, PtdIns*P*, and PtdInsP_2 and their lyso derivatives (see also Chapter 12 for an alternative HPLC method). Nakamura et al. (24) have reported an isocratic reversed phase HPLC separation and sensitive detection of PtdIns*P* and PtdInsP_2 as the mono and di-(9-anthroyl) derivatives, respectively. The derivatives are prepared by treating the dry PtdIns*P* and PtdInsP_2 with 9-anthroyldiazomethane in methanol. The components are eluted with dioxane/methanol/water/phosphoric acid (40:60:4:0.1, by vol.) and are detected by ultraviolet (UV) (254 nm), which is sufficiently sensitive to discern as little as 0.25 μg of polyphosphoinositide. Other glycerophospholipids do not react with this reagent or are not recovered under the above chromatographic conditions (e.g., phosphatidic acid, PtdOH).

Glycosyl Phosphatidylinositol

Of the inositol phosphatides, only PtdIns and its palmitylated derivative have been found thus far to be conjugated to protein molecules by covalent bonding (4,25). For analysis of these anchor lipids, the hydrophobic domains are prelabelled with the photoreactive reagent 3-(trifluoromethyl)-3-(*m*-[^{125}I]iodophenyl) diazirine ([^{125}I]TID, Amersham Corp.) using detergent-free aggregates of the protein. Papain digestion of the labelled protein releases the hydrophobic domain, which is purified (26). In case of the PtdIns glycans (prelabelled with [^{125}I]TID, the protein is treated with purified phospholipase C (*Staphylococcus aureus*), which is specific for PtdIns, and the released diradylglycerols are recovered. In the case of [^{125}I]TID-labelled acetylcholinesterase from human red blood cells, which is resistant to phospholipase C (27), a PtdOH moiety may be released with a phospholipase D purified from plasma (28). Alternatively, the diradylglycerol moieties are recovered by acetolysis (28). The purification of the released lipids is performed by TLC, where the radiolabel does not affect the mobility of polar lipids significantly (26–28). However, a non-polar derivative such as an alkylglycerol diacetate can be affected by the incorporation of a radiolabel marker (27).

Figure 1 depicts schematically the isolation of PtdIns, PtdIns*P* and PtdInsP_2 from natural sources using a combination of neutral and acidic solvents and chromatography on adsorbents. Depending on the requirements of the experimental protocol, the procedure can be abbreviated and the lipid classes isolated directly from the total lipid extracts. The schematic includes the recovery of the diradylglycerol moieties of the inositol phosphatides by hydrolysis with an appropriate phospholipase C.

Free Diradylglycerols

Diradylglycerols from resting-state and agonist-stimulated cell cultures maintained in serum-free medium are isolated by conventional methods (7,29,30). Cultures are incubated with fresh serum-free medium with and without the agonist. Incubations are terminated by adding 1–2 ml of ice-cold methanol to the culture media. The cells are scraped into this methanol and added to 1–2 ml of chloroform. The dishes are washed with an additional 1–2 ml of cold methanol, which is also added to the chloroform and the lipid extraction completed. The final organic phase from several dishes is combined and a portion taken for quantification by GLC (2,5), following trimethylsilylation, or by radio-TLC (31) after first reacting with radioactive acetic anhydride. The remaining combined organic phase is dried in a rotary evaporator and the diradylglycerols are isolated as a group by TLC on silica gel G with diethylether/hexane/acetic acid (70:30:1, by vol.) (29) or normal phase HPLC with hexane/isopropanol/acetic acid (100:1:0.01, by vol.) as eluting solvent (11). In order to remove the products of acyl migration, the diradylglycerols should finally be isolated by TLC on borate-treated silica gel with chloroform/methanol (96:4, v/v) as the developing solvent (32).

CHARACTERIZATION OF FATTY ACYL AND ALKYL CHAINS

The diradylglycerol moieties of the glycerophospholipids are made up of acyl and alkyl chains of various length and degree of unsaturation. They differ in positional distribution and assume characteristic molecular associations, which must be determined for identification of individual molecular species.

Overall Composition

The total fatty acid composition of PtdIns, PtdIns*P* and PtdInsP_2 is determined following a 2 h reaction at 80°C with fresh methanol/sulphuric acid

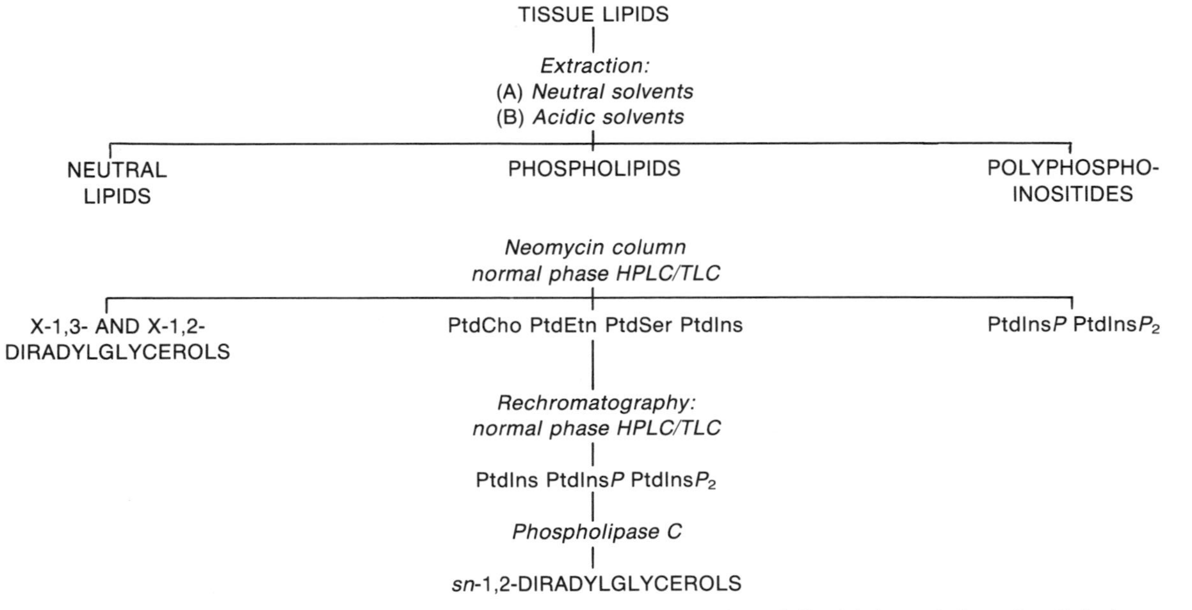

FIG. 1. Schematic representation of isolation, separation and preparation of diradylglycerols from inositol phosphatide classes. The free X-1,3- and X-1,2-diradylglycerols are isolated in parallel and are eventually processed similarly to the *sn*-1,2-diradylglycerols derived from the glycerophospholipids, as explained in text.

(94:6, v/v) (5), 15% (v/v) BF_3 in methanol (11), or a 10 min reaction at 20°C with 1 M $NaOCH_3$ in methanol/benzene (toluene (60:40, v/v)) (33). A highly sensitive measurement of fatty acids in the PtdIns*P* and PtdInsP_2, which are present at lower mass levels, has been obtained by UV absorption after preparation of the 2-nitrophenylhydrazide (3) or 9-anthryl (34) derivatives.

Positional Distribution of Fatty Acids

The positional distribution of the fatty acids in PtdIns is determined by hydrolysis with phospholipase A_2 (*Crotalus adamanteus*, Sigma Chemical Co.).

For this purpose (5) a portion of the purified phospholipid (0.5 mg) is vortexed for 3 h at 37°C in a mixture made up of 2 ml of buffer (17.5 mM tris(hydroxymethyl)aminomethane, adjusted to pH 7.3 with HCl, 1.0 mM $CaCl_2$), 2 ml diethyl ether, 50 μg of butylated hydroxytoluene (BHT) and 50 units of phospholipase A_2. After acidification with one drop of concentrated HCl, the mixture is extracted first with five parts chloroform/methanol (2:1, v/v), then with two parts chloroform/methanol (4:1, v/v). Two portions of the extract are then resolved by TLC on Silica Gel H. The plates are first developed to a height of 11 cm using chloroform/methanol/acetic acid/water (100:35:10:3, by vol.), dried under nitrogen gas for 10 min and then developed to a height of 15 cm using heptane/isopropyl ether /acetic acid (60:40:4, by vol.). The silica gel in the zones corresponding to free fatty acids, residual PtdIns and lyso PtdIns is scraped into tubes and then treated for 2 h at 80°C with 1.5 ml of methanol/sulphuric acid (94:6, v/v).

The fatty acid methyl esters and any dimethylacetals are extracted with hexane as follows.

After cooling of the tubes, 4 ml of hexane and 0.5 ml of chilled 3 M aqueous ammonia are added. After vortexing the extraction mixture, the tubes are centrifuged at 800 *g* for a few minutes. The hexane extract is then passed through a short column of anhydrous sodium sulphate in a Pasteur pipette containing a small plug of glass wool. (These columns must be prewashed with 2 ml of chloroform followed by 2 ml of hexane.) The hexane extract is evaporated under a stream of nitrogen gas. It is especially important with small samples to stop the evaporation process as soon as the solvent is no longer visible in the tube, otherwise it is possible to distort the composition by preferential evaporation of the shorter chain fatty acid methyl esters. Alkylglycerols are not analysed by this procedure. In order to measure the extent of hydrolysis one set of the tubes is quantified by GLC after addition of 50 μg of methyl heptadecanoate as internal standard. The residual PtdIns corresponds to approximately 0.5% of total in most digestions.

The PtdIns*P* and PtdInsP_2 can be hydrolysed to the corresponding lyso compounds by incubating with bee venom phospholipase A_2 (2), which has been purified (35). Alternatively, the positional distribution of the fatty acids can be determined following preparation of diacylglycerols by hydrolysis with pancreatic lipase (2), which attacks the primary positions of a diacylglycerol.

Distribution of Alkyl and Alkenyl Ether Groups

The fatty acid and alkylglycerol composition of the alkylacylglycerophospholipids is determined by GLC following chemical degradation. The composition of fatty acids can be determined on very small quantities of material provided sample contamination is avoided.

For this purpose (28,36) 1 M anhydrous methanolic HCl (100 μl) is added to the dried sample and the reaction mixture is heated at 65°C for 16 h. For analysis of fatty acid methyl esters only, samples are extracted with 2,2,4-trimethylpentane. For combined analysis of fatty acid methyl esters and alkylglycerols, 200 μl of chloroform and 75 μl of H_2O are added to the sample, and, after vortexing, the lower organic phase is removed. Samples for subsequent analysis by TLC are then dried. For GLC analysis, the aqueous phase is re-extracted with chloroform (100 μl), and the combined organic extracts are acetylated by incubation with 10 μl of acetic anhydride/pyridine (1:1, v/v) for 30 min at 80°C, dried, and the sample resuspended in 10 μl of 2,2,4-trimethylpentane.

The release of alkyl- and alk-1-enylglycerols from diradylglycerophospholipids can also be achieved by reduction with sodium bis(2-methoxyethoxy) aluminium hydride (Vitride™ or Red-Al™) (37,38). Compared to $LiAlH_4$, Vitride™ reagent gives better recoveries of alk-1-enylglycerols and can be purchased as a solution in benzene or toluene (Aldrich Chemical Co.). The resulting alkyl- and alk-1-enylglycerols along with the fatty alcohols are isolated by TLC with neutral lipid solvent systems and are converted to trimethylsilyl (TMS) ethers, acetates or isopropylidines for GLC analysis.

The fatty acid and alkylglycerol composition of alkylacylglycerol TMS ethers isolated by HPLC are determined following a basic methanolysis for 15 min with 0.5 ml $NaOCH_3$ (1 M in methanol/toluene (3:2, v/v), at 20°C. The mixture is then neutralized with 1% (v/v) acetic acid in hexane and extracted by adding 200 μl of chloroform, 0.5 ml of H_2O and 50 μl of 3 M aqueous ammonia. The organic phase is washed with 250 μl of H_2O and then dried by passing it through a small column of anhydrous sodium sulphate. After being dried under nitrogen gas, the sample is acetylated for 0.5 h at 80°C with acetic anhydride/pyridine (1:1, v/v). The reagents are removed by evaporation under nitrogen gas and the products purified by normal phase HPLC on a silica column with 0.8% (v/v) isopropanol in hexane (flow rate = 1 ml/min) (33,39). The fractions that correspond to standard fatty acid methyl esters (4.3–7.3 min) and alkylglycerol derivatives (7.3–10.6 min) are collected and evaporated to dryness. The residues are taken up in hexane

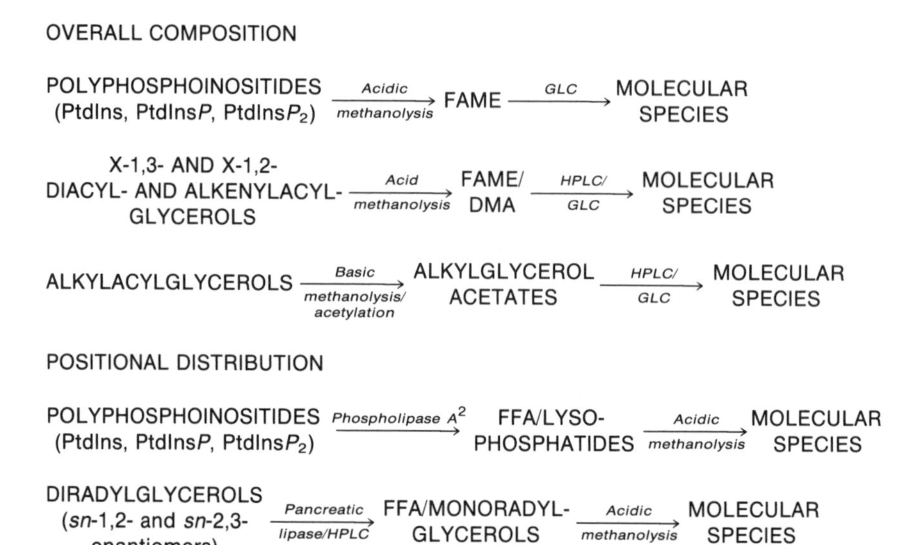

FIG. 2. Determination of fatty acyl chain composition and positional distribution. Some of the steps constitute parallel options, alternative methodology, or both, which may be omitted as required. These determinations are usually carried out on individual lipid classes, but they are more informative when performed on appropriate subfractions of the total or on individual molecular species. FAME, fatty acid methyl esters; DMA, dimethylacetals; FFA, free fatty acids.

and examined by GLC on both polar and non-polar capillary columns as described (33,36).

Figure 2 summarizes schematically the various methods of characterization of the overall composition and positional distribution of the radyl chains in the inositol phosphatides.

DERIVATIZATION AND SEPARATION OF SUBCLASSES

Although the molecular association of the radyl chains in the glycerophospholipids, including inositol phosphatides, can be determined directly by reversed phase HPLC and fast atom bombardment–mass spectrometry (FAB-MS) (see below), it is more convenient to assess it by examining the diradylglycerol moieties of the phospholipids. The diradylglycerols are readily released by phospholipase C and are converted into derivatives appropriate for chromatographic analysis. The conversion of diradylglycerols into derivatives prevents their isomerization and improves their chromatographic separation and detection. Furthermore, the derivative preparation and purification is well suited for a preliminary resolution of alkylacyl, alkenylacyl and diacylglycerol subclasses or of any enantiomeric diradylglycerols that may be present. This preliminary separation step further improves the resolution of molecular species and simplifies their identification.

Release of Diradylglycerols

The diradylglycerols are released from the parent glycerophospholipids by hydrolysis with phospholipase C (2). The enzyme can be either specific or non-specific for the phospholipid class. Cruder preparations of phospholipase C from *Bacillus cereus* (such as type III and some batches of type V, Sigma Chemical Co.) have activity against PtdIns and can be used to hydrolyse that phospholipid as well as phosphatidylcholine (PtdCho), phosphatidylethanolamine (PtdEtn) and phosphatidylserine (PtdSer). Purer preparations (such as type XIII, Sigma Chemical Co.) have no phospholipase C activity with PtdIns. The purified PtdIns-specific phospholipase C from *B. cereus* is also commecially available (Boehringer Mannheim Co.). The hydrolysis is illustrated with phospholipase C from *B. cereus* (Sigma Chemical Co., type III or type V), which yields an *sn*-1,2-diacylglycerol and a water-soluble inositol phosphate.

For this purpose (5) the purified PtdIns (approximately 500 μg) plus 50 μg of BHT as antioxidant is suspended in 1.5 ml of peroxide-free diethyl ether and 1.5 ml of buffer (17.5 mM tris(hydroxy-methyl)aminomethane adjusted to pH 7.3 with HCl, 1.0 mM $CaCl_2$), along with 10 units of the enzyme. The mixture is agitated using a vortex mixer for 10 s and then shaken at 37°C on a Buchler rotary Evapo-Mix for 3 h. The diradylglycerols are extracted two times with 2 ml of diethyl ether. The extracts are combined and passed through two successive Pasteur pipettes containing anhydrous Na_2SO_4. After reduction of volume, the extent of hydrolysis is checked by spotting the sample in a narrow lane (about 1 cm) beside 25 μg of undigested phospholipid on a silica gel plate, followed by developing in chloroform/methanol/aqueous ammonia (65:25:4, by vol.). After it has been sprayed with dichlorofluorescein, the amount of any residual phospholipid is estimated by a visual comparison with 25 μg of reference phospholipid standard.

Once it has been established that hydrolysis is at least 95% complete, further samples can be hydrolysed under the same conditions without further check. For analysis of molecular species the extracts containing the diradylglycerols are evaporated under nitrogen gas at room temperature. The samples are derivatized as soon as possible after removal of all solvent. If care is taken, there should be no detectable isomerization of 1,2- to 1,3-diacylglycerols.

The PtdIns glycans are hydrolysed using a PtdIns-specific phospholipase C from *S. aureus* (supplied by Dr M. Low) or *Bacillus thuringiens* (ICN Co.). The method is illustrated with a sample of purified bovine erythrocyte acetylcholine esterase (36).

For this purpose the protein (3–12 nmol) in 20 mM phosphate buffer, pH 7, sodium deoxycholate is treated for 90 min at 37°C with the enzyme. Diradylglycerols are extracted with three 1 ml portions of hexane.

Alternatively, acetolysis may be employed as a method of preparation of diradylglycerol acetates from PtdIns glycans that cannot be hydrolyzed with phospholipase C because of palmitylation of the inositol moiety (28). Such PtdIns glycans, however, are attacked by a specific phospholipase D (from plasma). This enzyme yields water-soluble inositol and phosphatidic acid, which can be resolved into molecular species by reversed phase HPLC following diazomethylation (40,41). Acetolysis may also be used for the preparation of diradylglycerol acetates from PtdIns*P* and PtdInsP_2, but it must be realized that this leads to migration of acyl groups (42). The mixture of 1,2- and 1,3-diacylglycerol acetates generated by acetolysis makes a detailed analysis of the molecular species more difficult, although the 1,2- and 1,3-forms can be purified before analysis by TLC. The analysis by GLC on polar capillary columns is also much slower with the acetate than with TMS derivatives.

Holub et al. (2) have described conditions for hydrolysis of PtdIns*P* and PtdInsP_2 with a PtdIns-specific phospholipase C from bovine brain. In a scaled-down version, the phosphatide (3 mg) could be incubated in 5 ml of 0.05 M Tris·HCl buffer (pH 7.2) containing brain extract (1 mg of protein) and 0.025% (w/v) cetyltrimethylammonium bromide at 37°C for 60 min. This enzyme also hydrolyses brain PtdIns. The liberated diacylglycerols are immediately removed from the incubation medium by extraction with diethyl ether and derivatized. The diradylglycerol acetates have been analysed by GLC on non-polar columns and were used for assessment of molecular species composition (2). The brain enzymes have now been purified from particulate (43) and cytosol (44) fractions.

Preparation of Derivatives of Diradylglycerols

For chromatographic separation into enantiomeric subfractions and molecular species, the diradylglycerols are converted into appropriate derivatives. For resolution of enantiomers on chiral phase HPLC columns, the diradylglycerols are converted into dinitrophenylurethanes (45).

For this purpose, diradylglycerol samples of less than 1 mg are treated with about 2 mg of 3,5-dinitrophenylisocyanate (Sumitomo Chemical Co., Osaka, Japan) in 4 ml of dry toluene in the presence of dry pyridine (40 μl) at ambient temperature for 1 h. The crude urethane derivatives of diacyl and dialkylglycerols are purified by TLC on Silica Gel GF plates (20 cm × 20 cm × 0.25 mm, Analtech Inc., Newark, DE) using hexane/ethylene dichloride/ethanol (40:10:3 and 40:10:1, by vol., respectively). Prior to use, the plates are activated by heating at 110–120°C for 2 h. The dinitrophenylurethanes of the diacylglycerols (R_f = 0.52) and dialkylglycerols (R_f = 0.63) are recovered by eluting the gel scrapings with diethyl ether (45). Recently, dinitrophenylurethane (DNPU) derivatives of enantiomeric alkylacylglycerols have been prepared (46).

For resolution of enantiomers on normal phase HPLC columns, the diradylglycerols are converted to the corresponding 1-(1-naphthyl)ethyl urethane diastereomers (47).

For this purpose, the diacylglycerols (1–2 mg) are dissolved in dry toluene (300 μl) and treated with the (R)- or (S)-forms of 1-(1-naphthyl)ethyl isocyanate (10 μl) in the presence of 4-pyrrolidinopyridine (approximately 10 μmol) overnight at 50°C. The products are extracted with hexane/diethyl ether (1:1, v/v) and washed with 2 M HCl and H_2O. The organic layer is taken to dryness under a stream of nitrogen gas and the sample is purified on a short column of Florisil eluted with ether.

The DNPU derivatives can be cleaved to recover the original diradylglycerols without isomerization (48), but the naphthylethyl urethanes have not been examined in this reaction. The cleavage of the DNPU derivatives of diradylglycerols is accomplished with trichlorosilane under the general conditions described by Pirkle and Hauske (49).

For this purpose (48) 200–400 μg of DNPU is dissolved in 400 μl of dry toluene, plus 40 μl each of triethylamine/dry toluene (1:20, v/v) and trichlorosilane/dry toluene (1:20, v/v). After the mixture has stood for 16 h at 21–22°C, two drops of H_2O are added and the mixture is vortexed vigorously for 30 s. The products are extracted with 5 ml of diethyl ether, dried over anhydrous Na_2SO_4 and evaporated under nitrogen gas.

The diradylglycerols are converted into TMS or t-butyldimethylsilyl (TBDMS) ethers by reaction with the corresponding trialkylchlorosilane under conditions that do not lead to isomerization during the reaction.

Thus, the TMS ethers are prepared (50) by reaction with 100 μl of pyridine/hexamethyldisilazane/trimethylchlorosilane (15:5:2, by vol.) for 30 min at room temperature. The reagents are then evaporated under nitrogen gas and the products dissolved in 2 ml of hexane. After brief centrifugation (1000 *g*), the supernatant is evaporated under nitrogen gas and the sample dissolved in an appropriate volume of hexane. The centrifugation step can be avoided by using pyridine/bis(trimethylsilyl)-trifluoroacetamide/trimethylchlorosilane (50:49:1, by vol.), which yields volatile by-products. The TBDMS ethers are prepared (51) by reacting the diradylglycerols with 150 μl of a solution of 1 M t-butyldimethylchlorosilane and 2.5 M imidazole in dimethylformamide for 20 min at 80°C. After the reaction mixture is dissolved in 5 ml of hexane, the products are washed three times with H_2O and then dried by passage through a small column containing anhydrous sodium sulphate.

For HPLC analysis with UV detection, the diradylglycerols are converted into the benzoates (52), dinitrobenzoates (34,53), or pentafluorobenzoates (54) by reaction with the corresponding acid anhydride or acid chloride and pyridine for 10–15 min at 60–80°C (53). The pentafluorobenzoates yield intense negative ions in chemical ionization mass spectrometry and are well suited for combined liquid chromatography–mass spectrometry (LC-MS) applications (54). For HPLC analysis with fluorescence detection, the diradylglycerols are converted into naphthylurethanes (55) or 1-anthroyl derivatives (56). The preparation of these derivatives requires somewhat higher temperatures and longer times of reaction than the preparation of the silyl ethers of benzoyl esters.

For preparation of the benzoates (52) up to 2 mg of diradylglycerols are dissolved in 0.3 ml of benzene containing 10 mg of benzoic anhydride and 4 mg of 4-dimethylaminopyridine and are allowed to stand for 1 h. The samples are placed in an ice bath and 2 ml of 0.1 M NaOH added slowly. The diradylglycerobenzoates produced are extracted three times with hexane. When the amount of diradylglycerols reaches 10 mg, the benzoylation reagents are increased four-fold.

For preparation of dinitrobenzoates (53) 2 mg of diacylglycerol and 25 mg of dinitrobenzoyl chloride are dried for 30 min *in vacuo* prior to use. The mixture is dissolved in 0.5 ml of dry pyridine and heated in a sealed vial at 60°C for 10 min. After cooling, the reaction is stopped by adding 2 ml of 80% (v/v) methanol and 2 ml of H_2O. Then the reaction mixture is applied to a SepPak C_{18} cartridge (Waters Associates, Walton, MA) and the cartridge is washed with 25 ml of 80% methanol to purify the sample. The product is recovered by eluting with 30 ml of methanol. After evaporation, the residue is dissolved in hexane and washed with H_2O. Diethyl ether and methanol contained 0.01% BHT.

For preparation of the naphthylurethanes (55), the diradylglycerols (200 nmol) are dissolved in 100 μl of dimethylformamide. Alpha-naphthylisocyanate (10 μl) yielding at least 500-fold excess of the reagent is then added along with 1,4-diazabicyclo(2,2,2)octane (0.1 M in dimethylformamide). The stoppered vial is heated at 85°C for 30 min. The excess reagent is destroyed by addition of 300 μl of methanol and the reaction mixture centrifuged to obtain a clear supernatant, which is evaporated to a final volume of 100 μl.

PtdIns may be subjected directly to derivatization methods, which preserve both the polar and the non-polar parts in the same molecule. This permits partial resolution of the molecular species on silica gel impregnated with silver nitrate. For this purpose the inositol ring of PtdIns is cleaved first with periodate, and the oxidation products are then treated with diazomethane (57). The dimethylphosphatidic acid produced is purified by preparative TLC and resolved into molecular species on the basis of degree of unsaturation by $AgNO_3$-TLC (58). The molecular species present in the different TLC bands are identified by determining the fatty acid composition. This method is not widely used, but could be of interest again because of the discovery of a phospholipase D, that is specific for PtdIns (59). However, reversed phase HPLC (40,41) would now be preferred to $AgNO_3$-TLC for the resolution of the molecular species.

Separation of Diradylglycerol Subclasses

Since the various diradylglycerophospholipids cannot be resolved into their subclasses as the intact phosphatides, this must be accomplished following derivatization. The separation of the diacylglycerol (R_f = 0.42) and alkenylacylglycerol (R_f = 0.61) subclasses can be accomplished in the free form by TLC on Silica Gel G containing 5% (w/v) boric acid with chloroform/acetone (97.5:2.5, v/v) as the developing solvent (32). BHT runs to the solvent front in this system. The TBDMS ethers of alkenylacylglycerols (R_f = 0.66) and diacylglycerols (R_f = 0.35) are also well resolved by TLC on Silica Gel H with benzene (or toluene) as developing solvent (60). BHT has an R_f = 0.82 in this system. When present in significant amounts, the alkylacylglycerols migrate to a position just above the diacylglycerols in this solvent system and it is difficult to obtain a pure alkylacylglycerol fraction, when the diacylglycerol fraction is present in larger quantities. A somewhat better TLC resolution of the diacyl, alkylacyl and alkenylacylglycerol subclasses is obtained following acetylation (61). Since HPLC offers higher resolution and is also more appropriate for smaller samples, the diradylglycerols are best resolved by HPLC.

The diradylglycerol TMS ethers are resolved into subclasses by normal phase HPLC on a silica column (Supelco, Inc., Bellefonte, PA) with 0.3% (v/v) isopropanol, in hexane (flow rate = 1 ml/min) (39). Fractions corresponding to reference alkenylacyl (3.8–4.8 min), alkylacyl (4.8–6.3 min) and diacyl (6.3–8.5 min) glycerol TMS ethers are collected and evaporated to dryness. The residues are taken up in hexane and appropriate portions examined by polar and non-polar capillary GLC with flame ionization detection or by GC–MS. Similar HPLC separations of the subclasses of the diradylglycerols are obtained for the TBDMS ethers (62), as well as for the benzoyl (52), dinitrobenzoyl (53), naphthylurethane (55) and 1-anthroyl (56) derivatives. Thus, the separation of the alkylacyl and diacylglycerol naphthylurethanes (55) is obtained on a Lichrosorb Si100 column (250 mm × 4 mm) with hexane/tetrahydrofuran (95:5, v/v) as the eluting solvent (flow rate = 1 ml/min). The alkylacyl (12–15 min) and the diacyl derivatives (22–27 min) were clearly resolved from each other but partly contaminated with UV-absorbing unknowns, which, however, did not contain significant amounts of fatty acids.

It is possible that the DNPU derivatives of diradylglycerols could also be separated into subclasses by normal phase HPLC, but the exact experimental conditions have not been established. Takagi and Itabashi (45) have shown, however, that from chiral phase HPLC columns the dialkylglycerol derivatives emerge well ahead of the diacylglycerol derivatives using the OA-4100 stationary phase and hexane/ethylene dichloride/ethanol (80:20:1, by vol.) as the mobile phase. It would be anticipated that the alkenylacylglycerol DNPU derivatives would be eluted ahead of the dialkylglycerol derivatives. The resolution of the diradylglycerols into the individual enantiomer subclasses is best performed following a preliminary separation of the diradylglycerols into the alkylacyl and diacylglycerol subclasses either at the free diradylglycerol stage or following the preparation of the TMS ethers. The recovered TMS ethers of the diradylglycerol subclasses can be converted into the DNPU derivatives without isomerization. For this purpose, the TMS ethers are reacted directly with the 3,5-dinitrophenylisocyanate or after a prior hydrolysis with H_2O.

A partial resolution of the TMS, TBDMS, acetyl and benzoyl derivatives of the diradylglycerol subclasses is also obtained by GLC on non-polar capillary columns, which are used for quality control of all chromatographic and chemical manipulations during the diradylglycerol preparations. However, very limited resolution beyond simple carbon numbers is achieved (63,64).

The TBDMS ethers (51) and the benzoyl esters (52) of each diradylglycerol subclass may be prefractionated according to the number of double bonds by TLC on Silica Gel G made up with 20% (w/v) silver nitrate. A two-stage development is used in order to resolve the polyunsaturated species as well as the oligo-enoic species (50). The plate is first developed to a distance of 10 cm with chloroform/methanol (94:6, v/v) and then, after the plate has been dried 10 min under a stream of nitrogen gas, to a height of 16 cm, with chloroform alone. The TBDMS ethers are eluted from the gel with chloroform/methanol (2:1, v/v). After washing with dilute aqueous ammonia, the extracts are dried by passing through a column containing anhydrous sodium sulphate and then taken to dryness. The contents of each band were analysed by GLC and GC-MS (51). Portions may also be subjected to acidic or basic methanolysis and TLC separation and the resulting fatty acid methyl esters, dimethylacetals or alkylglycerols analysed by GLC.

Figure 3 summarizes the various transformations employed to obtain a complete identification of the molecular species of diradylglycerols including their enantiomeric form. Direct evidence of the reverse isomer composition of each species, however, cannot be obtained by present chromatographic methods, but mass spectrometry shows promise (51).

RESOLUTION OF MOLECULAR SPECIES OF DIRADYLGLYCEROLS

The common molecular species of diradylglycerols are most extensively resolved by GLC on polar capillary columns. Reversed phase HPLC is less efficient, but offers advantages for the separation of polyunsaturated species. HPLC also allows peak collection. Chiral phase HPLC provides only a limited resolution of molecular species, but ensures their enantiomeric purity. For peak identification each chromatographic system can be effectively em-

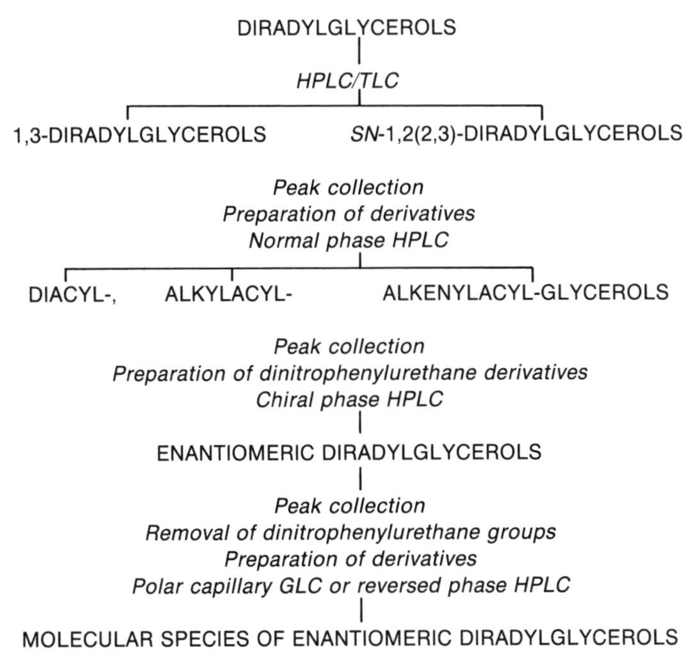

FIG. 3. Schematic representation of separation and identification of molecular species of diradylglycerols. Some of the steps may be carried out in parallel, while the order of others may be reversed, with still others omitted altogether.

ployed in combination with mass spectrometry, although this is expensive and frequently unnecessary.

Polar Capillary GLC and GC-MS

Non-polar capillary columns provide a highly sensitive and reliable separation and quantification of molecular species within each diradylglycerol subclass on the basis of molecular weight or carbon number. The TMS ethers of the diradylglycerols are resolved on an 8 m fused silica capillary column coated with a non-polar permanently bonded SE-54 liquid phase. The samples are injected on-column and the oven temperature is programmed from 40 to 150°C at 30 deg.C/min, then to 230°C at 20 deg.C/min, and to 340°C at 5 deg.C/min (65). Hydrogen at 6 p.s.i. (1 p.s.i. $\approx$ 6.9 kPa) head pressure is used as carrier gas. This separation may be combined with a preliminary resolution of the diacylglycerols by $AgNO_3$-TLC, $AgNO_3$-HPLC or reversed phase HPLC (66).

GLC on polar capillary columns provides the most extensive resolution of molecular species yet achieved by any method within each class of diradylglycerols. The molecular species are resolved on the basis of both carbon and double-bond number by GLC at 250°C on a 10 m capillary column coated with an SP-2330 polar liquid phase (68% cyanopropyl, 32% phenylsiloxane) supplied by Supelco (66). The temperatures of the injector and detector are maintained at 270°C and 300°C, respectively. The samples are introduced with a split injector (split ratio 7:1) and hydrogen is used as the carrier gas. More recently (67,68), diradylglycerol TMS ethers have been separated on fused silica columns (15 m × 0.32 mm) coated with cross-bonded RTx 2330 (Restek Corp., Port Matilda, PA) or with stabilized SP2380 (Supelco) (68,69). The carrier gas was hydrogen at 3 p.s.i. head pressure and the column temperature isothermal at 250°C. The sample was admitted by a split injector (split ratio 7:1). Determinations of major components (greater than 10%) had coefficients of variation (CV) of 2% or less, whereas minor components (less than 1%) had CV of approximately 10%. Figure 4 shows the resolution of molecular species of the diacylglycerol moieties of human red blood cell PtdIns (69) on polar and non-polar capillary columns. While the non-polar GLC column (b) yields only a few peaks, the polar capillary column (a) resolves 10 times more components. Figure 5 shows the resolution of the molecular species of the alkylacylglycerol moieties of the PtdIns anchor of human red blood cell acetylcholinesterase on polar (a) and non-polar (b) capillary columns (28).

The diradylglycerol peaks of the various chromatograms are usually identified in relation to the retention times of either standards or well-characterized natural samples (68). This is generally satisfactory for many mixtures, where the fatty acid and alkyl and alkenylglycerol composition of the sample is also determined. When these cannot be measured, which may be the case for very small or extremely complex samples, relative retention times are not sufficient for reliable peak identification. In such instances combinations of GLC or HPLC with mass spectrometry are helpful. Since polar capillary columns have low loading capacity and give too much bleed, thermally stable non-polar columns are usually employed for GC-MS identification of molecular species of diradylglycerols, although such columns provide only carbon number resolution.

Thus, the TBDMS ethers of diacylglycerols are identified by GC-MS using a non-polar column, which separates the molecular species on the basis of carbon number (51). The species within each GLC peak or carbon number are then identified on the basis of the pseudomolecular $[M - 57]^+$ ion, the $[\text{acyl} + 74]^+$ ions, and the fragment ions formed by loss of one of the acyl chains $[\text{R} - \text{RCOO}]^+$. The $[M - 57]^+$ ions arise from a facile loss of the t-butyl group, which constitutes a special advantage in the identification of molecular species of diradylglycerols by mass spectrometry. Figure 6 illustrates the identification of the TBDMS ethers of a diacylglycerol mixture derived from plasma high-density lipoprotein phosphatidylcholine (PtdCho) by non-polar capillary GLC and mass spectrometry (63). The relatively

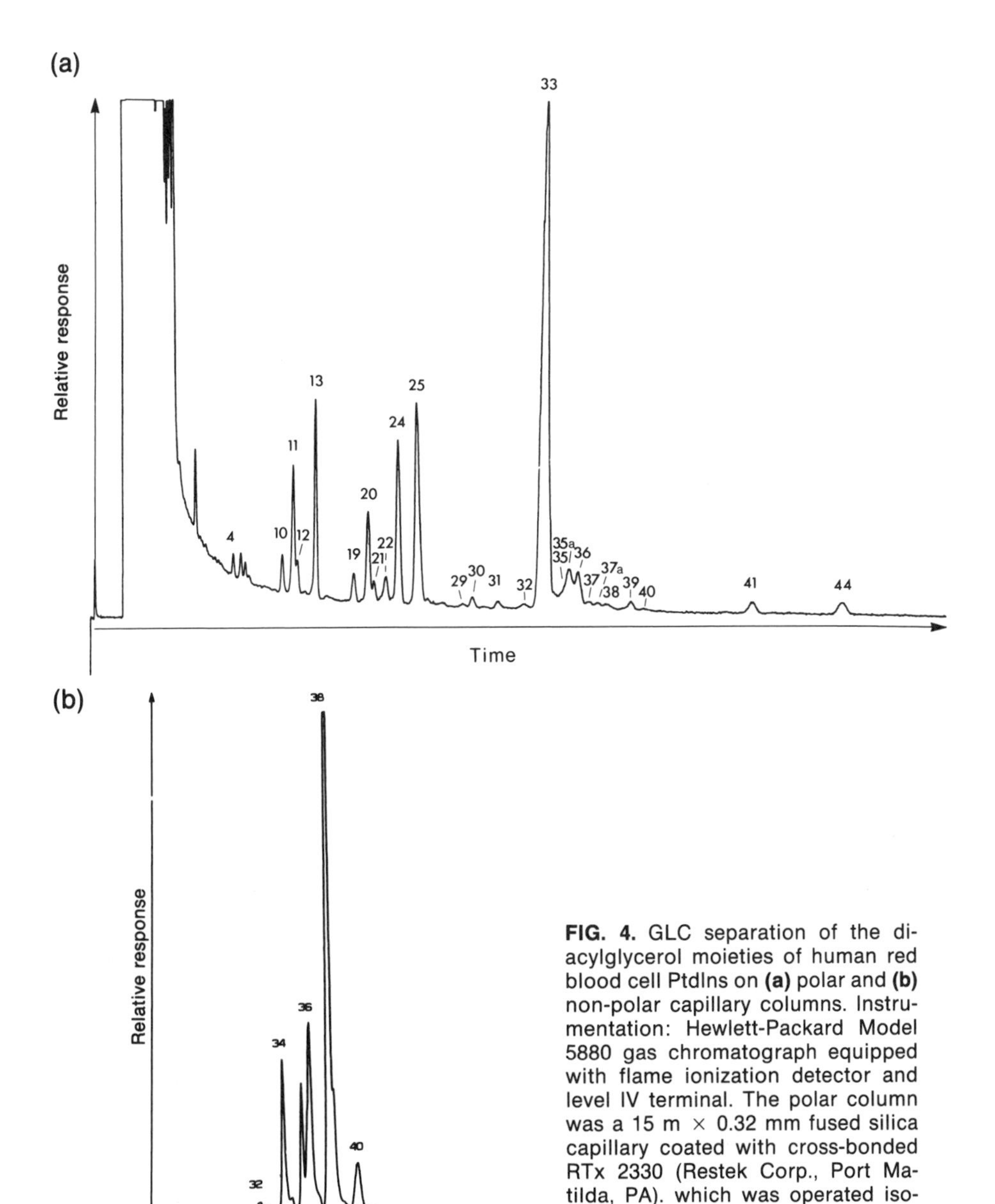

FIG. 4. GLC separation of the diacylglycerol moieties of human red blood cell PtdIns on **(a)** polar and **(b)** non-polar capillary columns. Instrumentation: Hewlett-Packard Model 5880 gas chromatograph equipped with flame ionization detector and level IV terminal. The polar column was a 15 m × 0.32 mm fused silica capillary coated with cross-bonded RTx 2330 (Restek Corp., Port Matilda, PA). which was operated isothermally (250°C) with hydrogen as carrier gas (3 p.s.i. head pressure). The sample was introduced by split injection (split ratio 7:1). The non-polar column was an 8 m × 0.32 mm fused silica capillary coated with cross-linked 5% phenylmethyl silicone (Hewlett-Packard). The sample was injected on-column and temperature was programmed in four steps from 40 to 350°C as given in text. The carrier gas was hydrogen at 6 p.s.i. head pressure. Sample: 1 μl of 0.1% diacylglycerol TMS ethers in hexane.

Peak identification: (a) 4, 16:0–16:0; 10, 16:0–18:0; 11, 16:0–18:1(9); 12, 16:0–18:1(7); 13, 16:0–18:2; 19, 18:0–18:0; 20, 18:0–18:1(9); 21, 18:0–18:!(7); 22, 18:1–18:1(9)(9); 24, 18:0–18:2; 25, 16:0–20:4 plus 16:0–20:3 plus 18:1(9)–18:2; 29, 18:2–18:2 plus 16:0–20:5; 30, 17:0–20:4; 33, 18:0–20:4(6) plus 18:0–20:3(6); 35, 16:0–22:4(6); 35a, 18:1–20:3 plus 18:1 plus –20:4; 36, 18:1(9)–20:4; 37, 18:1(7)–20:4; 37a, unknown; 38, 18:0–20:5(3); 39, 16:0–22:5(3) plus 16:0–22:6(3); 40, 18:2–20:4(6); 41, 18:0–22:4(6); 44, 18:0–22:5 plus 18:0–22:6. (b) Peaks identified by total number of acyl carbons.

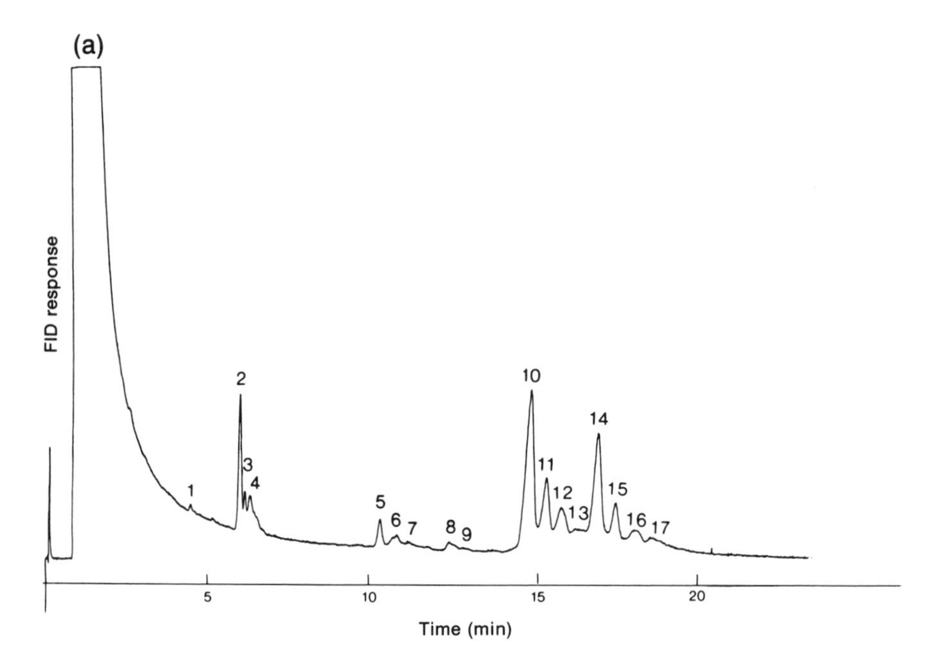

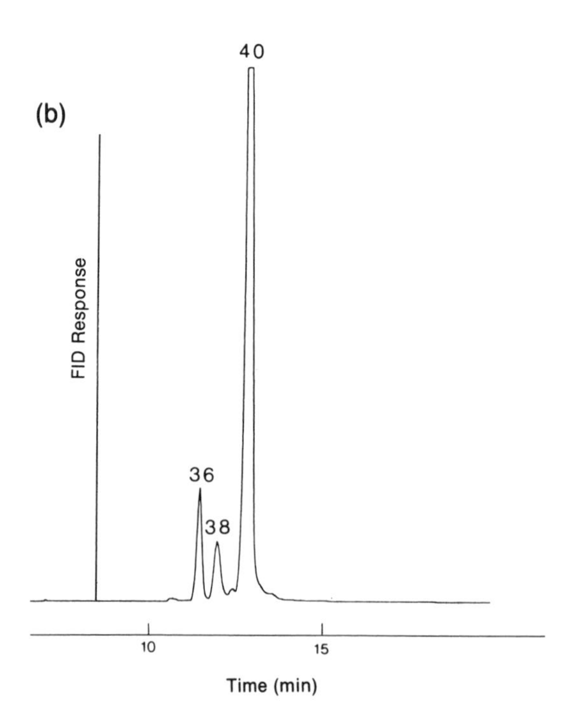

FIG. 5. GLC separation of the alkylacylglycerol moieties of the PtdIns anchor of acetylcholinesterase from human red blood cells on (a) polar and (b) nonpolar capillary columns (28). Peak identification: 1, 34:0; 2, 1,2–18:0′–18:0; 3, 1,3–18:0′–18:0; 4, 18:1′–18:0; 5, 1,2–18:0′–20:4; 6, 1,3–18:0′–20:4; 7, unidentified; 8, 1,2–17:0′–22:4; 9, 1,3–17:0′–22:4; 10, 1,2–18:0′–22:4; 11, 1,3–18:0′–22:4; 12, 1,2–18:1′–22:4; 13, 1,3–18:1′–22:4; 14, 1,2–18:0′–22:5 plus 18:0′–22:6; 15, 1,3–18:0′–22:5 plus 18:0′–22:6; 16, 1,2–18:1′–22:5 plus 18:1′–22:6; 17, 1,3–18:1′–22:5 plus 18:1′–22:6. A prime indicates an alkyl group. Peaks 36, 38 and 40, diradylglycerols with a combined alkyl and acyl carbon number of 36, 38 and 40. Instrumentation and analytical conditions as given in Fig. 4. Sample: 1 μl of approximately 0.01% diradylglycerol acetates in hexane.

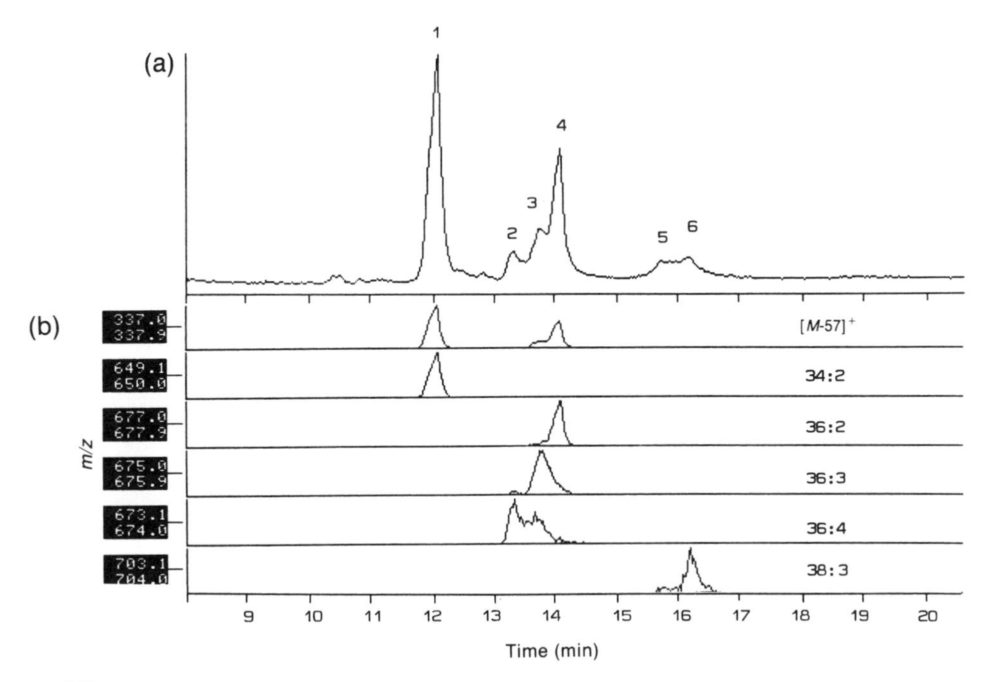

FIG. 6. Identification of plasma high-density lipoprotein PtdCho diacylglycerol TBDMS ethers by GC-MS (63). (a) Total ion current; (b) mass chromatograms of single ions $[M - 57]^+$ representing the major diradylglycerol species. Instrumentation: Hewlett-Packard Model 5880 gas chromatograph equipped with a non-polar flexible quartz capillary column (8 m × 0.30 mm internal diameter) coated with a permanently bonded SE-54 liquid phase (Hewlett-Packard) coupled to a Hewlett-Packard quadrupole mass spectrometer, Model 5985B. GLC conditions as in Fig. 1 for a non-polar capillary column. Mass spectrometric conditions: electron impact, 70 eV. Mass range 200–700 Da. Peak identification: 1, 16:0–18:2 plus 16:0–18:1; 2, 16:0–20:4; 3, 18:1–18:2 plus 18:2–18:2; 4, 18:0–18:1 plus 18:0–18:2; 5, 16:0–22:6 plus 18:0–20:4; 6, 18:0–20:3. m/z 337, [RCO + 74] corresponding to 18:2.

efficient capillary column allows some resolution of the molecular species within each carbon number.

Reversed Phase HPLC and LC-MS

Alternatively, the molecular species of the diradylglycerols can be resolved by reversed phase HPLC. This is especially convenient or if UV-absorbing or fluorescent derivatives are prepared. The benzoyl (52,70) and dinitrobenzoyl (53,71) derivatives have been extensively used for the resolution of molecular species of diradylglycerols from a variety of glycerophospholipids. Thus, the diacyl-, alkenylacyl- and alkylacylglycerol moieties of the choline phosphatides have been resolved as the benzoates by reversed phase

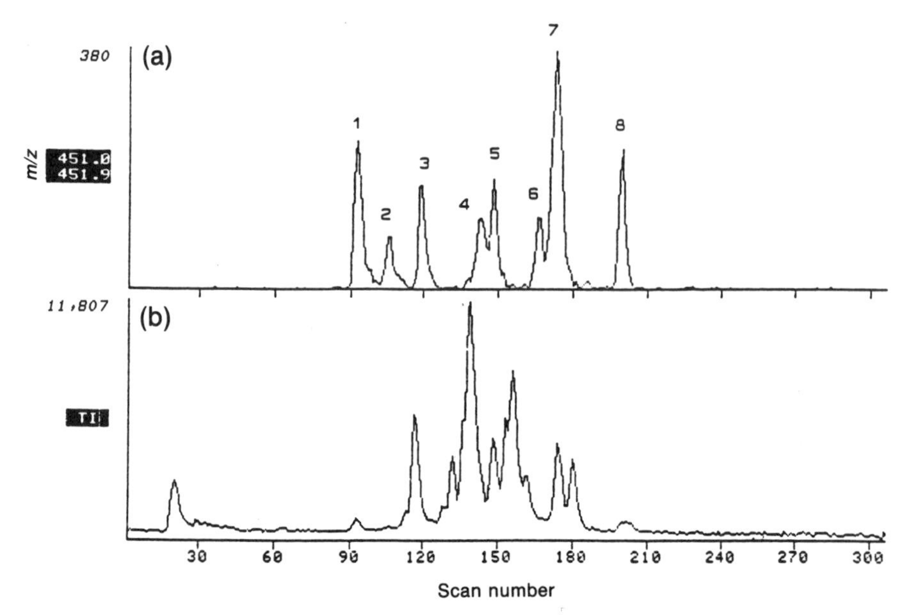

FIG. 7. LC-CIMS of rat heart PtdEtn diradylglycerol TBDMS mixture (73). **(a)** Mass chromatograms extracted from the total ion current spectrum **(b)** showing all alkenyl (identified below by ″) species containing 18:2 fatty acid. Instrumentation: liquid chromatograph Model 1084B coupled to a quadrupole mass spectrometer Model 5985 via a direct liquid inlet interface (Hewlett-Packard, Palo Alto, CA). HPLC conditions: 30–90% propionitrile in acetonitrile over 30 min. Positive chemical ionization. Peak identification: 1, 18:2–22:6; 2, 18:2–20:4; 3, 18:2–18:2; 4, 18:1–18:2; 5, 16:0–18:2; 6, 18:1″–18:2; 7, 16:0″–18:2; 8, 18:0″–18:2. Here and in subsequent figures, italic numbers denote total number of ions.

HPLC using a 25 cm × 0.46 cm (internal diameter) Ultrasphere ODS columns (5 μm packing) and solvents made up of acetonitrile/isopropanol in the ratios 70:30, 65:35 and 63:37 (v/v), respectively, with a flow rate of 1.0 ml/min (52). Haroldsen and Murphy (71) used desorption electron capture negative chemical ionization (NCI) to identify the dinitrobenzoyl esters of the diacyl- and alkenylacylglycerol molecular species that were separated by reversed phase HPLC on an Ultrasphere ODS column (5 μm, 25 cm × 0.45 cm) and a 5 μm ODS precolumn cartridge (Waters Associates) with the solvent system acetonitrile/isopropanol (85:15, v/v) and an on-line UV monitor (254 nm).

The TBDMS ethers of the diradylglycerols are stable to moisture and can be effectively employed in reversed phase liquid chromatography–chemical ionization–mass spectrometry (LC-CI-MS), where the facile loss of the t-butyldimethylsilanol group $[MH - 132]^+$ together with the formation of prominent $[MH - RCOOH]^+$ ions provides the information necessary for peak identification of the molecular species (72). Figure 7 illustrates the application of LC-CI-MS to the identification of the diradylglycerol moieties of rat heart phosphatidyl ethanolamine (PtdEtn) as the TBDMS ethers

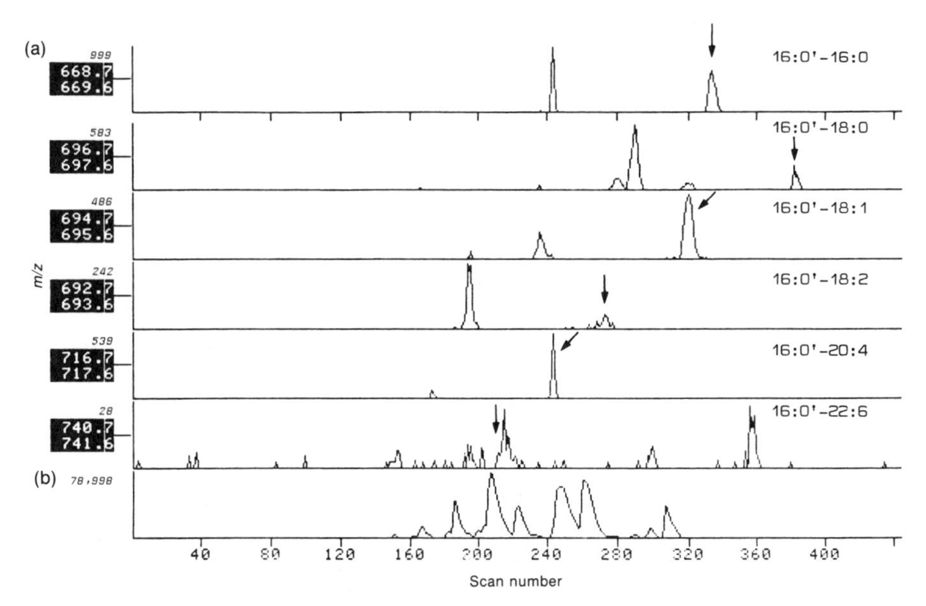

FIG. 8. LC-CI-MS of diradylglycerol moieties of rat red blood cell PtdCho (74). (a) Mass chromatograms of $[M - 57]^+$ ions containing 16:0′ (alkyl) as one of the radyl chains; (b) total ion current response. Instrumentation and LC-MS conditions as given in Fig. 7.

(73,74). The mass chromatograms extracted from the total ion current spectrum identify all the species containing 18:2 fatty acid, among which are many alkenylacylglycerol species, as indicated in the figure. The peak identification is based on both the characteristic *m*/*z* values and the relative HPLC elution times. Figure 8 provides a similar illustration of the application of LC-CI-MS to the identification of alkylacylglycerol species among the diradylglycerol moieties derived from rat red blood cell PtdCho (74). Again the peaks have been identified on the basis of the characteristic *m*/*z* values and the relative HPLC retention time.

Reversed phase LC-CI-MS is also well suited for the identification of benzoyl and pentafluorobenzoyl derivatives of diradylglycerols (74). The strong UV absorption of these derivatives allows an effective comparison of the relative quantitative responses of the UV and the CI-MS signals. Of special interest is the reversed phase LC-MS of the pentafluorobenzoates, which yield a very strong signal for the molecular ion in NCI. The signal for the $[M]^-$ ion in the NCI is 500–1000 times stronger than any signal in positive chemical ionization (PCI). The NCI spectrum, however, does not yield any information about the structural components of the molecular species of the diradylglycerols, which must be obtained from the relative elution times and by positive CI, where the $[MH - RCOOH]^+$ ions predominate, although at much reduced overall sensitivity, when compared to the $[M]^-$ ion in NCI.

Chiral Phase HPLC and LC-MS

In addition to the resolution of enantiomers, chiral phase HPLC also yields a limited separation of molecular species. In any event the resolved peaks can be collected and the radyl chain composition determined. Alternatively, the DNPU groups can be removed without isomerization using trichlorosilane and the resulting diradylglycerols reanalysed as the TMS ethers on polar capillary GLC (48).

However, chiral phase HPLC can also be combined with mass spectrometry and the molecular species making up any one of the chromatographic peaks identified on the basis of their molecular weight and characteristic fragment ions (75). The DNPU derivatives of diradylglycerols yield informative mass spectra in negative chemical ionization with chloride attachment (76).

Chiral phase LC-MS allows the identification of the individual molecular species of the diradylglycerols following the chromatographic resolution of the enantiomers (75). The chiral columns provide complete resolution of enantiomers containing the same radyl chains, but mixtures of enantiomers containing different radyl chains can lead to overlaps among different species of the *sn*-1,2- and *sn*-2,3-enantiomers. These overlaps may be readily identified by chiral phase LC-MS of the diradylglycerol DNPU derivatives. Figure 9 illustrates the detection and quantification of free diacylglycerols from rat intestinal microsomes by means of chiral phase LC-MS with chloride attachment NCI (77). It shows that the acylation of 2-mono-oleoylglycerol by a monoacylglycerol acyltransferase leads to the formation of both *sn*-1,2- and *sn*-2,3-dioleoylglycerols, although the *sn*-1,2-enantiomer is present in a much larger proportion (90% of total). The more sensitive detection of the DNPU derivatives by chloride-attachment NCI mass spectrometry is readily obtained under chiral phase LC-MS conditions by including about 1% (v/v) methylene chloride in the eluting solvent. The molecular species are identified by the prominent $[\text{diacylglycerol} + \text{Cl}]^-$ ions, the yields of which are comparable among the species of different degrees of unsaturation within a single carbon number or a small group of carbon numbers (78). In contrast, the yields of $[M + \text{Cl}]^-$ are small and much more variable. Similar methods could be employed in the determination of the chiral structure and molecular identity of other diacylglycerols in tissue extracts of incubation mixtures.

Although LC-MS has not been employed previously for the identification of the molecular species of diacylglycerols resolved on normal phase HPLC following preparation of the diastereomeric naphthylethylurethanes (47), direct probe analysis has shown that these compounds possess excellent mass spectrometric properties (79).

RESOLUTION OF INTACT INOSITOL PHOSPHATIDES

Although less efficient and more complicated, analyses of molecular species of intact phosphatidylinositols are possible and necessary when assessing

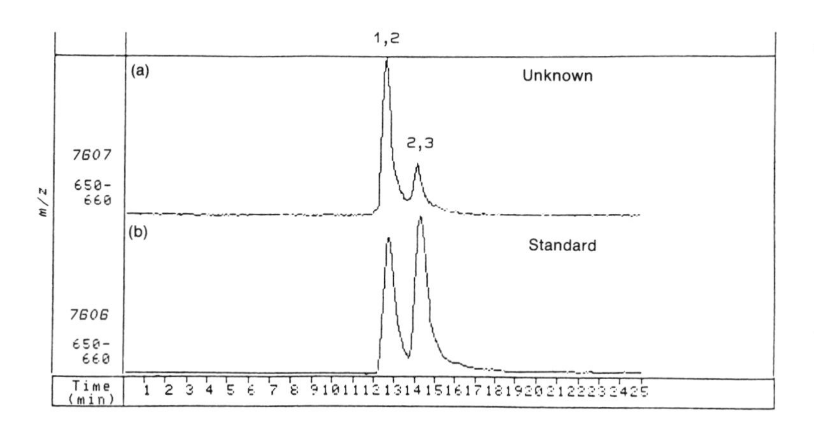

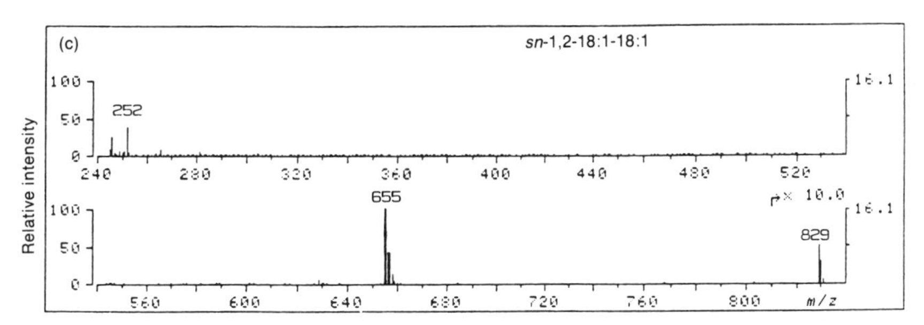

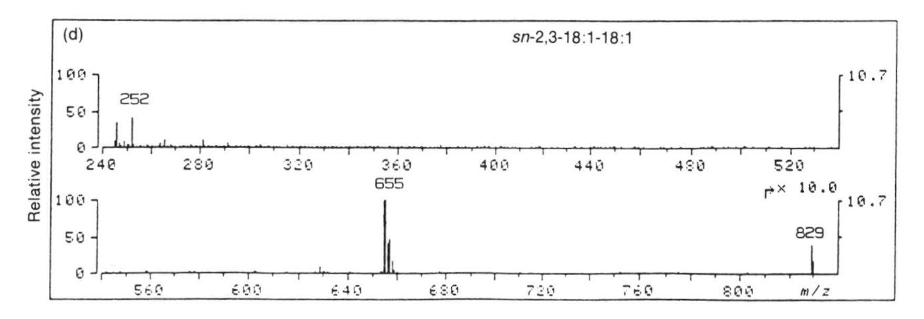

FIG. 9. Chiral phase LC-MS detection and quantification of *sn*-2,3-dioleoylglycerol among the free diacylglycerol products of microsomal acylation of 2-mono-oleoylglycerol (77). **(a)** Total ion current of unknown sample (mass range 650–660 Da); **(b)** total ion current of standard (mass range 650–660 Da); **(c)** full mass spectrum of peak 1,2 of unknown sample; **(d)** full mass spectrum of peak 2,3 of unknown sample. Instrumentation as in Fig. 7. HPLC conditions: column, 25 cm × 0.46 cm internal diameter; chiral phase, YMC-Pack A-KO3 (YMC Inc., Kyoto, Japan); eluting solvent, isocratic iso-octane/t-butylmethyl ether/ isopropanol/acetonitrile (80:10:5:5, by vol.) containing 1% dichloroethane.

the molecular association of the radyl chains with radio- or stable-isotope-labelled polar head groups. These analyses are best performed with reversed phase HPLC or FAB-MS.

Reversed Phase HPLC and LC-MS

Reversed phase HPLC of intact PtdIns can be performed without prior derivatization of the molecule, although it complicates detection and quantification (6,70). Thus, excellent separations of the molecular species of PtdIns from rat liver have been obtained on a 250 mm × 4.6 mm column of 5 μm particle size (Ultrasphere ODS). The column is eluted with 20 mM choline chloride in methanol/water/acetonitrile (90.5:7:2.5, by vol.) (6). The eluate is monitored at 205 nm. For quantification, however, it is necessary to calibrate the UV response, which varies greatly with the unsaturation of the molecular species. Figure 10 shows the resolution of the molecular species of intact PtdIns from bovine liver by a reversed phase LC-MS with a ther-

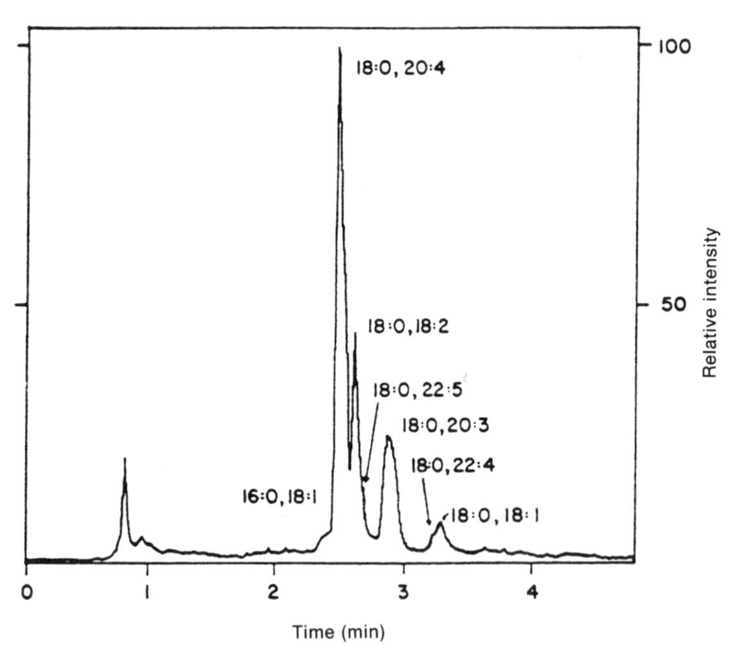

FIG. 10. Reversed phase LC-MS profile of the PtdIns of rat liver. Instrumentation: Beckman Model 114M liquid chromatograph equipped with a Du Pont Zorbax C-18 column (5 μm, 4.6 mm × 25 cm) and connected to an Extrel 400-2 quadrupole mass spectrometer via a Vestec thermospray interface (Vestec, Houston, TX). Total ion chromatogram (mass range 105–1005 Da) obtained for 20 μg of phosphatidylinositol from bovine liver using methanol/hexane/0.1 M NH_4OAc (500:25:25, by vol.) as developing solvent. (Redrawn, with permission, from Ref. 80.)

mospray interface (80). The column is eluted with methanol/hexane/0.1 M ammonium acetate in H_2O (500:25:25, by vol.) with a flow rate of 1 ml/min or 2 ml/min. Under the thermospray conditions, the spectrum of a PtdIns species is dominated by monoacyl- ($[RCO + 74]^+$), and diacylglycerol-type fragments. There is also an ion due to the ammonium adduct of inositol at *m*/*z* 198.

FAB-MS

Finally, it must be pointed out that the molecular species of diradylglycerols can be resolved, identified and quantified by FAB-MS of the intact original glycerophospholipids, including the phosphatidylinositols. Thus, Sherman et al. (81) have obtained useful spectra of the PtdIns from soybean, and the PtdIns(4)*P* and PtdIns(4,5)*P* from bovine brain. The mass spectra were obtained on a VG ZAB-HF operated as a double focusing instrument with 8 keV energy and about 1000 resolution. Mass analysed ion kinetic spectra (MIKES) were obtained on the same instrument. FAB ionization was performed with xenon atoms using the gun arrangement normally fitted by VG. Samples were applied directly into 2–3 μl of glycerol on the sample holder. It was shown that analysis on a triple sector mass spectrometer with and without collisional activation was necessary for complete composition information especially when the parent ion contains isobaric species. Analysis of the $[M - H]^-$ ions for fatty ester composition by means of MIKES was not adequate to resolve fatty acyl daughter ions when the parent ion contains isobaric species. In both instances quantitative analysis of the fatty ester content of individual molecular species was complicated by dissimilar ion yields from fatty acyl-bearing fragments from compositionally different parent ions (81).

FAB-MS has also been utilized in the identification of the glycosyl inositide backbone and molecular species of the deacylated glycosyl inositides from the lipid anchor of human red blood cell acetylcholine esterase (82).

SUMMARY AND CONCLUSIONS

A review of the methods employed for mass analyses of diradylglycerols from resting and agonist-stimulated cells or tissues reveals a great variety of experimental approaches. The methods range from crude assays of fatty acids or radioactivity in arbitrarily chosen sections of thin-layer plates, developed without the benefit of appropriate reference standards, to highly sophisticated routines involving the isolation and complete structural characterization and quantification of individual molecular species. Clearly, the latter approach is not practical or even necessary in analysis of serial samples. However, the analyte composition must be established before shortcuts are used to simplify the analysis of large numbers of samples. This chapter provides an extensive compilation of detailed analytical methods from which appropriate experimental routines can be selected as required by the specific protocol of the study. The discussion recognizes that the basic difficulty in these mass analyses is the paucity rather than the complexity of the diradylglycerol samples. Therefore, contamination of the small amounts of diradylglycerol lipid by other tissue lipids and by lipid contaminants in reagents and solvents is a major problem, which cannot be eliminated entirely. Likewise, serious consideration must be given to isomerization, peroxidation and degradation during the processing of the diradylglycerols, which may be rich in polyunsaturated fatty acids. Since antioxidants are resolved from diradylglycerols during chromatography, addition of these agents is effective only in preventing oxidation during the storage.

After the diradylglycerol sample has been procured, it is essential to monitor the products of further transformations and fractionations. Due to its high sensitivity, capillary GLC is ideally suited for this purpose, allowing the use of only a small fraction of the sample. HPTLC may also be sufficiently sensitive in many cases, although it lacks the advantage of easy quantification by flame ionization. These methods are applicable to the original diradylglycerol mixture and to most derivatives. The choice of a particular routine must be decided on the basis of clearly defined but limited experimental needs, although an absence of more complete analyses may be regretted afterwards. Alternatively, if the quantity permits, the sample may be subjected to parallel analyses via two or more analytical routines to obtain other types of information. A complete analysis on a single sample may not be practical, although it can be outlined theoretically.

Normal phase HPLC is well suited for the resolution of the alkenylacyl-, alkylacyl- and diacylglycerol subclasses. These group separations are important because of the well-recognized diversity in the origin of the diradylglycerols, including both glycerophospholipids and triacylglycerols, as well as both lipid anchor and free inositol phosphatides. Chiral phase HPLC is essential to distinguish between enantiomers of diradylglycerols. A clear differentiation between the enantiomers is of special importance in analysis of the free diradylglycerol fraction, since the resting species may include a high proportion of products from triacylglycerol lipolysis, which may yield both *sn*-1,2- and *sn*-2,3-isomers. GLC on polar capillary columns provides the most extensive resolution of molecular species of the diradylglycerol subclasses, although difficulties may be experienced with the recovery of species containing two polyunsaturated fatty chains. These species are best resolved and recovered by reversed phase HPLC. Detailed knowledge of the structure and quantitative composition of molecular species of the diradylglycerols is necessary to establish the origin and function of these compounds as substrates of enzymatic transformations and as cellular messengers.

Acknowledgments The studies by the authors and their collaborators were supported by funds from the Medical Research Council of Canada, Ottawa, Ontario, the Heart and Stroke Foundation of Ontario, Toronto, Ontario, and the Hospital for Sick Children Foundation, Toronto, Ontario. Appreciation is expressed to Dr W. W. Christie for providing a preprint of a paper in press.

REFERENCES

1. Holub BJ, Kuksis A. Metabolism of molecular species of diacylglycerophospholipids. *Adv Lipid Res* 1978;16:1–125.
2. Holub BJ, Kuksis A, Thompson W. Molecular species of mono-, di- and triphosphoinositides of bovine brain. *J Lipid Res* 1970;11:558–564.
3. Augert G, Blackmore PF, Exton JH. Changes in the concentration and fatty acid composition of phosphoinositides induced by hormones in hepatocytes. *J Biol Chem* 1989;264:2574–2580.
4. Ferguson MAJ, Williams AF. Cell surface anchoring in proteins via glycosyl-phosphatidylinositol structures. *Annu Rev Biochem* 1988;57:285–320.
5. Myher JJ, Kuksis A. Molecular species of plant phosphatidylinositol with selective cytotoxicity towards tumor cells. *Biochim Biophys Acta* 1984;795:85–90.
6. Patton GM, Fasulo JM, Robins SJ. Separation of phospholipids and individual molecular species of phospholipids by high performance liquid chromatography. *J Lipid Res* 1982;23:190–196.
7. Pessin MS, Raben DM. Molecular species analysis of 1,2-diglycerides stimulated by alpha-thrombin in cultured fibroblasts. *J Biol Chem* 1989;264:8727–8738.
8. Berridge MJ. Cell signaling through phospholipid metabolism. *J Cell Sci (Suppl)* 1986;4:137–153.
9. Fisher SK, Agranoff BW. Receptor activation and inositol lipid hydrolysis in neural tissues. *J Neurochem* 1987;48:999–1017.
10. Exton JH. Signaling through phosphatidylcholine breakdown. *J Biol Chem* 1990;265:1–4.
11. Bocckino SB, Blackmore PF, Exton JH. Stimulation of 1,2-diacylglycerol accumulation in hepatocytes by vasopressin, epinephrine, and angiotensin II. *J Biol Chem* 1985;260:14,201–14,207.
12. Kennerly DA. Diacylglycerol metabolism in mast cells. *J Biol Chem* 1987;262:16,305–16,313.
13. Nishizuka Y. The molecular heterogeneity of protein kinase C and its implications for cellular regulation. *Nature* 1988;334:661–665.
14. Boni LT, Rando RR. The nature of protein kinase C activation by physically defined phospholipid vesicles and diacylglycerols. *J Biol Chem* 1985;260:10,819–10,825.
15. Abdel-Latif AA. Calcium-mobilizing receptors, polyphosphoinositides, and the generation of second messengers. *Pharmacol Revs* 1986;38:227–272.
16. Shaikh NA. Phospholipid analysis. In: Fozzard HA, Haber E, Jennings RB, Katz A, et al., eds. *The heart and cardiovascular system*, vol. 1. New York: Raven Press, 1986;289–302.
17. Schacht J. Purification of polyphosphoinositides by chromatography on immobilized neomycin. *J Lipid Res* 1978;19:1063–1067.
18. Palmer FBStC. Chromatography of acidic phospholipids on immobilized neomycin. *J Lipid Res* 1981;22:1296–1300.
19. Jolles J, Zwiers H, Dekker A, et al. Corticotropin-(1-24)-tetracosapeptide affects protein phosphorylation and polyphosphoinositide metabolism in rat brain. *Biochem J* 1981;194:283–291.
20. Akhtar RA, Taft WC, Abdel-Latif AA. Effects of ACTH on polyphosphoinositide metab-

olism and protein phosphorylation in rabbit iris subcellular fractions. *J Neurochem* 1983;41:1460–1468.
21. Abdel-Latif AA, Akhtar RA, Hawthorne JN. Acetylcholine increases the breakdown of triphosphoinositide of rabbit iris muscle prelabelled with [^{32}P]phosphate. *Biochem J* 1977;162:61–73.
22. Fukami K, Takenawa T. Quantitative changes in polyphosphoinositides, 1,2-diacylglycerol, and inositol 1,4,5-triphosphate by platelet-derived growth factor and prostaglandin F_{2alpha}. *J Biol Chem* 1989;264:14,985–14,989.
23. Boyle LE, Sklar LA, Traynor-Kaplan AE. An HPLC procedure for separating polyphosphoinositides on hydroxylapatite. *J Lipid Res* 1990;31:157–159.
24. Nakamura T, Hatori Y, Yamada K, et al. A high performance liquid chromatographic method for the determination of polyphosphoinositides in brain. *Anal Biochem* 1989;179:127–130.
25. Roberts WL, Rosenberry TL. Identification of covalently attached fatty acids in the hydrophobic membrane-binding domain of human erythrocyte acetylcholinesterase. *Biochim Biophys Res Commun* 1985;133:621–627.
26. Roberts WL, Rosenberry TL. Selective radiolabeling and isolation of the hydrophobic membrane-binding domain of human erythrocyte acetylcholinesterase. *Biochemistry* 1986;25:3091–3098.
27. Roberts WL, Kim BH, Rosenberry TL. Differences in the glycolipid membrane anchors of bovine and human erythrocyte acetylcholinesterases. *Proc Natl Acad Sci USA* 1987;84:7817–7821.
28. Roberts WL, Myher JJ, Kuksis A, et al. Lipid analysis of the glycoinositol phospholipid membrane anchor of human erythrocyte acetylcholinesterase. *J Biol Chem* 1988; 263:18,766–18,775.
29. Takuwa N, Takuwa Y, Rasmussen H. A tumour promoter, 12-*O*-tetradecanoylphorbol 13-acetate, increases cellular 1,2-diacylglycerol content through a mechanism other than polyphosphoinositide hydrolysis in Swiss-mouse 3T3 fibroblasts. *Biochem J* 1987;243:647–653.
30. Wright TRM, Raugan LA, Shaw HS, et al. Kinetic analysis of 1,2-diacylglycerol mass levels in cultured fibroblasts. *J Biol Chem* 1988;263:9374–9380.
31. Hoskins JA, Evans CE. Measurement of intracellular diacylglycerols by acetylation and thin-layer chromatography. *J Chromatogr Biomed Applic* 1989;490:439–443.
32. Thomas AE III, Scharoun JE, Ralston H. Quantitative estimation of isomeric monoglycerides by thin-layer chromatography. *J Am Oil Chem Soc* 1965;42:789–792.
33. Myher JJ, Kuksis A, Pind S. Molecular species of glycerophospholipids and sphingomyelins of human erythrocytes: Improved method of analysis. *Lipids* 1989;24:396–407.
34. Takamura H, Tanaka K, Matsuura T, et al. Ether phospholipid molecular species in human platelets. *J Biochem* 1989;105:168–172.
35. Tsai TC, Hart H, Jiang RT, et al. Phospholipids chiral at phosphorus. Use of chiral thiolphosphatidylcholine to study the metal binding properties of bee venom phospholipase A_2. *B iochemistry* 1985;24:3180–3188.
36. Roberts WL, Myher JJ, Kuksis A, et al. Alkylacylglycerol molecular species in the glycosylinositol phospholipid membrane anchor of bovine erythrocyte acetylcholinesterase. *Biochem Biophys Res Commun* 1988;150:271–277.
37. Snyder F, Blank ML, Wykle RL. The enzymatic synthesis of ethanolamine plasmalogens. *J Biol Chem* 1971;246:3639–3645.
38. Schmid HHO. Bandi PC, Kwei LS. Analysis and quantitation of ether lipids by chromatographic methods. *J Chromatogr Sci* 1975;13:478–486.
39. Kuksis A, Myher JJ. Analytical methodology in the fat absorption area. In: Kuksis A, ed. *Fat absorption*, vol. 1. Boca Raton, FL: CRC Press, 1986;1–141.
40. Nakagawa Y, Fujishima K, Waku K. Separation of dimethylphosphatidates of alkylglycerophosphocholine and their molecular species by high performance liquid chromatography. *Anal Biochem* 1986;157:172–178.
41. Nakagawa Y, Waku K. Improved procedure for the separation of the molecular species of dimethylphosphatidate by high performance liquid chromatography. *J Chromatogr* 1986;381:225–231.
42. Privett OS, Nutter LJ. Determination of the structure of lecithins via the formation of acetylated 1,2-diglycerides. *Lipids* 1977;2:149–154.

43. Katan M, Parker PJ. Purification of phosphoinositide-specific phospholipase C from a particulate fraction of bovine brain. *Eur J Biochem* 1987;168:413–418.
44. Ryu SH, Cho KS, Lee KY, et al. Two forms of phosphatidylinositol-specific phospholipase C from bovine brain. *Biochem Biophys Res Commun* 1986;141:137–144.
45. Takagi T, Itabashi Y. Rapid separation of diacyl- and dialkylglycerol enantiomers by high performance liquid chromatography on a chiral stationary phase. *Lipids* 1987;22:596–600.
46. Takagi T, Okamoto J, Ando Y, et al. Separation of the enantiomers of 1-alkyl-2-acyl-*rac*-glycerol and of 1-alkyl-3-acyl-*rac*-glycerol by high performance liquid chromatography on a chiral column. *Lipids* 1990;25:108–110.
47. Laakso P, Christie WW. Chromatographic resolution of chiral diacylglycerol derivatives: potential in the stereospecific analysis of triacyl-sn-glycerols. *Lipids* 1990 (submitted).
48. Itabashi Y, Kuksis A, Myher JJ. Determination of molecular species of enantiomeric diacylglycerols by chiral phase high performance liquid chromatography and polar capillary gas chromatography. 1990. To be submitted.
49. Pirkle WH, Hauske JR. Trichlorosilane-induced cleavage. A mild method for retrieving carbinols from carbamates. *J Org Chem* 1977;42:2781–2782.
50. Myher JJ, Kuksis A. Improved resolution of natural diacylglycerols by gas–liquid chromatography on polar siloxanes. *J Chromatogr Sci* 1975;13:138–145.
51. Myher JJ, Kuksis A, Marai L, et al. Microdetermination of molecular species of oligo- and polyunsaturated diacylglycerols by gas chromatography–mass spectrometry of their tert-butyl dimethylsilyl ethers. *Anal Chem* 1978;50:557–561.
52. Blank ML, Robinson M, Fitzgerald V, et al. Novel quantitative method for determination of molecular species of phospholipids and diglycerides. *J Chromatogr* 1984;298:473–482.
53. Takamura H, Narita H, Urade R, et al. Quantitative analysis of polyenoic phospholipid molecular species by high performance liquid chromatography. *Lipids* 1986;21:356–361.
54. Kuksis A, Marai L, Myher JJ, Itabashi Y, Pind S. Applications of GC/MS, LC/MS, and FAB/MS to determination of molecular species of glycerolipids. In: *AOCS monograph on lipid and lipoprotein analysis*, Champaign: AOCS Press 1990 (in press).
55. Rabe H, Reichmann G, Nakagawa, Y, et al. Separation of alkylacyl- and diacyl glycerophospholipids and their molecular species as napthylurethanes by HPLC. *J Chromatogr Biomed Applic* 1989;353–360.
56. Ramesha CS, Pickett WC, Murthy DVK. Sensitive method for the analysis of phospholipid subclasses and molecular species as 1-anthroyl derivatives of their diglycerides. *J Chromatogr Biomed Applic* 1989;491:37–48.
57. Luthra MG, Sheltawy A. The fractionation of phosphatidylinositol into molecular species by thin-layer chromatography on silver nitrate-impregnated silica gel. *Biochem J* 1972;126:1231–1239.
58. Luthra MG, Sheltawy A. The metabolic turnover of molecular species of phosphatidylinositol and its precursor phosphatidic acid in guinea-pig cerebral hemispheres. *J Neurochem* 1976;27:1503–1511.
59. Low MG. Glycosyl-phosphatidylinositol: a versatile anchor for cell surface proteins. *Fed Am Soc Exp Biol* 1989;3:1600–1608.
60. Myher JJ, Kuksis A. Resolution of diacylglycerol moieties of natural glycerophospholipids by gas–liquid chromatography on polar capillary columns. *Can J Biochem Cell Biol* 1984;62:352–362.
61. Nakagawa Y, Sugiura T, Waku K. The molecular species composition of diacyl, alkylacyl and alkenylacylglycerophospholipids in rabbit alveolar macrophages. High amounts of 1-*O*-hexadecyl-2-arachidonoyl molecular species in alkylacylglycerophosphocholine. *Biochim Biophys Acta* 1985;833:323–329.
62. Myher JJ, Kuksis A, Pind S. Lipid composition of brush border membranes and microsomes of rat jejunal villus cells. *INFORM* 1990;1:332 [Abstract.]
63. Kuksis A, Myher JJ. New approaches to lipid analysis of lipoproteins and cell membranes. In: Kates M, Kuksis A, eds. *Membrane fluidity*, Clifton, NJ: Humana Press, 1980;3–32.
64. Traitler H. Recent advances in capillary gas chromatography applied to lipid analysis. *Progr Lipid Res* 1987;26:257–280.
65. Myher JJ, Kuksis A. Determination of plasma total lipid profiles by capillary gas–liquid chromatography. *J Biochem Biophys Methods* 1984;10:13–23.
66. Myher JJ, Kuksis A. Resolution of diacylglycerol moieties of natural glycerophospholipids

by gas–liquid chromatography on polar capillary columns. *Can J Biochem* 1982;60:638–650.

67. Myher JJ, Kuksis A, Pind S. Molecular species of glycerophospholipids and sphingomyelins of human erythrocytes: Improved method of analysis. *Lipids* 1989;24:396–407.
68. Myher JJ, Kuksis A. Relative gas–liquid chromatographic retention factors of trimethylsilyl ethers of diradylglycerols on polar capillary columns. *J Chromatogr* 1989;471:187–204.
69. Myher JJ, Kuksis A, Pind S. Molecular species of glycerophospholipids and sphingomyelins of human plasma: Comparison to red blood cells. *Lipids* 1989;24:408–418.
70. Robins SJ, Patton GM. Separation of phospholipid molecular species by high performance liquid chromatography: potentials for use in metabolic studies. *J Lipid Res* 1986;27:131–139.
71. Haroldsen PE, Murphy RC. Analysis of phospholipid molecular species in rat lung as dinitrobenzoate diglycerides by electron capture negative chemical ionization mass spectrometry. *Biomed Environ Mass Spectrom* 1987;14:573–578.
72. Pind S, Kuksis A, Myher JJ, et al. Resolution and quantitation of diacylglycerol moieties of natural glycerophospholipids by reversed-phase liquid chromatography with direct liquid inlet mass spectrometry. *Can J Biochem* 1984;62:301–309.
73. Kuksis A, Myher JJ. General strategies for practical chromatographic analysis of lipids. In: Kuksis A, ed. *Chromatography of lipids in biomedical research and clinical diagnosis*, Amsterdam: Elsevier, 1987;1–47.
74. Kuksis A, Marai L, Myher JJ, et al. Molecular speciation of natural glycerolipids and glycerophospholipids by liquid chromatography with mass spectrometry. In: Shukla VKS, Holmer G, eds. *Proceedings 15th scandinavian symposium on lipids*. Rebild Bakker, Denmark: Lipidforum, 1989;336–370.
75. Itabashi Y, Marai L, Kuksis A. Identification of molecular species of enantiomeric diacylglycerols by chiral phase LC/CIMS of the 3,5-dinitrophenylurethanes. *J Am Oil Chem Soc* 1989;66:491. [Abstract.]
76. Marai L, Kuksis A, Myher JJ, Itabashi Y. Liquid chromatography–chloride attachment negative chemical ionization mass spectrometry of neutral lipids. 1990; (to be submitted).
77. Lehner R, Itabashi Y, Kuksis A. Chiral column reassessment of the stereospecificity of monoacylglycerol acyltransferase from rat intestine. *INFORM* 1990;1:353 [Abstract.]
78. Kuksis A, Marai L, Myher JJ. Plasma lipid profiling using reversed phase HPLC with chloride-attachment negative chemical ionization/mass spectrometry. *J Am Oil Chem Soc* 1990 (in press).
79. Michelsen P, Aronsson E, Odham G, et al. Diastereomeric separation of natural glycerolderivatives as their 1-(1-naphthyl)ethyl carbamates by high performance liquid chromatography. *J Chromatogr* 1985;350:417–426.
80. Kim HY, Salem N Jr. Application of thermospray high performance liquid chromatography/mass spectrometry for the determination of phospholipids and related compounds. *Anal Chem* 1987;59:722–726.
81. Sherman WR, Ackermann KE, Bateman RH, et al. Mass-analyzed ion kinetic energy spectra and B_1E-B_2 triple sector mass spectrometric analysis of phosphoinositides by fast atom bombardment. *Biomed Mass Spectrom* 1985;12:409–413.
82. Roberts WL, Santikarn S, Reinhold VN, et al. Structural characterization of the glycoinositol phospholipid membrane anchor of human erythrocyte acetylcholinesterase by fast atom bombardment mass spectrometry. *J Biol Chem* 1988;263:18,776–18,784.

Subject Index

Explanations for abbreviations may be found on pp. xv–xvii